William Hargreaves

Our Wasted Resources

The Missing Link in the Temperance Reform

William Hargreaves

Our Wasted Resources
The Missing Link in the Temperance Reform

ISBN/EAN: 9783337296599

Printed in Europe, USA, Canada, Australia, Japan

Cover: Foto ©berggeist007 / pixelio.de

More available books at **www.hansebooks.com**

OUR WASTED RESOURCES;

THE

MISSING LINK IN THE TEMPERANCE REFORM.

BY

WILLIAM HARGREAVES, M.D.

————— ◆ ◆ ◆ —————

NEW YORK:

The National Temperance Society and Publication House,

No. 58 READE STREET.

——

1878.

INTRODUCTION.

It is vitally important that a people should encourage and foster only those trades and habits which tend to promote industry, sobriety, and virtue; for society is only safe while its members are industrious, sober, and virtuous.

Men do not generally wish to do what will either injure themselves or others; yet it must be admitted that society is virtually destroying itself by allowing habits and practices that always tend to produce unhappiness, idleness, vice, crime, and disease. For these evils are as truly the manufactured products of society as are boots, shoes, cotton goods, and broadcloth.

It is therefore clearly the interest and the duty of every citizen to regulate his own conduct and habits, that they may not only preserve his own honor and integrity, but that he may help to establish and extend like virtues throughout society.

The wise and good should endeavor to teach the masses whatever will contribute to their happiness; and the government should aim to control the passions of the vicious and inconsiderate members of society by such laws and regulations as will best tend to produce those results.

But laws, to be respected and generally obeyed, must commend themselves to, and obtain the approbation of, good men. Laws, to do this, must favor natural justice and tend to promote sobriety, industry, morality, and religion.

In the opinion of a very large and intelligent class of persons of the United States, Great Britain, and elsewhere, it is a very fatal and deplorable mistake for any government to license or allow any of its citizens to manufacture or to sell intoxicating liquors as a beverage; and that the only safe and true policy would be to prohibit both their manufacture and sale as common drinks.

Another class, perhaps the more numerous, advocate the license system as now generally practised in most States in the Union, as well as in Great Britain and many other countries.

The relation of the use and the traffic in strong

drinks to the trade, labor, and the general prosperity of the country is much more important than is generally conceived, and of which very little has been said or written by political economists.

It has been our aim in this essay to present such facts and figures on the subject treated as are reliable, in order that every one could judge intelligently. We do not desire to be deceived nor to deceive. The figures presented are official, or, if not, reasons are given why they are presented. The statistics of agriculture, manufactures, etc., given in the tables, and also in the comparisons made, are either the exact or the proportional amounts as found in Volume III. (Wealth and Industries) of the ninth census of the United States of 1870, or from other official reports. These facts and figures, and the deductions drawn from them, are respectfully presented to the serious consideration of my fellow-citizens and others who deem them worthy their consideration. And though several years have been chiefly spent in their preparation, I shall esteem the time not spent in vain, should they be a means of exciting a wider and a still deeper investigation of the effects of the use and sale of strong drinks on the industry, trade, and com-

merce of the country. Such investigations cannot fail to point out the true policy to be adopted
by the people in relation to the liquor-traffic and
the use of intoxicating drinks as a beverage.

PHILADELPHIA, PA., 1875.

CONTENTS.

10 CONTENTS.

TABLES.

OUR WASTED RESOURCES.

CHAPTER I.

NATIONAL WEALTH.

THE principal sources of national wealth are : I. Agriculture ; II. Manufactures ; III. Trade and Commerce ; IV. Railways ; V. Mines ; VI. Fisheries.

AGRICULTURAL RESOURCES.

The agricultural resources of the United States are unequalled by any other civilized nation.

The area of the United States is about three million square miles, three-fourths of which is capable of being inhabited.

By the census returns of 1870 there were, in farms, 407,735,041 acres, of which 188,921,099 acres were improved and 218,813,942 acres unimproved. The value of these farms was $9,262,803,861 ; and of farm implements and machinery, $336,878,429.*

The total value of "farm productions, betterments,
    and additions to stock" was $2,447,538,658
"Wages paid during the year, including value of
    board," 310,286,285

* Ninth Census Report, Vol. III., Wealth and Industries, pp. 81, 82.

13

" Value of animals slaughtered or sold for
    slaughter," $398,956,376
Value of " home manufactures," 23,423,332
Value of "forest productions," 36,808,277
Value of " market-garden productions," . . 20,719,229
Value of orchard products, ... 47,335,189

The magnitude of these figures, representing the
value of our agricultural products, gives but a very
slight idea of the immensity of the products of the
soil. It is only after much thought, comparison,
and analytic reasoning on the subject that the mind
can grasp or appreciate the magnitude of our agricultural resources.

The quantity of the principal cereals or breadstuffs
raised in the United States in 1870 was:

	Bushels.			Bushels.
Wheat,. . .	287,745,626	Oats, . . .		282,107,157
Rye, . . .	16,918,795	Barley,	. .	29,761,305
Indian corn, .	760,944,549	Buckwheat, .	.	9,821,721

Total breadstuffs, ...   1,387,299,153

This amount of breadstuffs raised in one year would
furnish 183 bushels to each of the 7,579,363 families,
or give 35 bushels to every man, woman, and child in
our country. The wheat alone is 7½ bushels per
head.

The principal fibrous productions raised in 1870
were : *

	Pounds.			Pounds.
Cotton,	. .	1,204,798,400	Hemp, . .	28,551,040
Wool,	. .	100,102,387	Silk cocoons, .	3,937
Flax, .	. .	27,133,034		

Total fibrous products, ... 1,360,588,798

* Ibid., p. 85.

Hay, hops, rice, and tobacco raised in 1870 were : *

Hay (in tons),	. 27,316,048	Rice (pounds),	. 73,635,021
Hops (in pounds), .	25,456,669	Tobacco (pounds), .	262,735,341

Potatoes, peas, beans, beeswax, honey, and domestic wine produced in 1870 were : *

	Bushels.		Pounds.
Irish potatoes,	. 143,337,473	Beeswax, . .	631,129
Sweet potatoes,	. 21,709,824	Bees' honey, . .	14,702,815
Peas and beans,	. 5,746,027	Dom'tic wine (gals.)	3,092,330

The sugar and molasses produced in 1870 were :

Cane sugar (hhds.),	87,043	Cane molasses (gals.),	6,593,323
Maple " (pounds),	28,443,645	Maple " "	921,057
Sorghum " (hhds.),	24	Sorghum " "	16,050,089

The amount of dairy products in 1870 was : *

Butter,	. . . . .	514,092,683 pounds.
Cheese,	. . . . .	53,492,153 "
Milk sold,	. . . .	235,500,599 gallons.

Of live stock on farms in 1870 there were : *

Horses,	. . . 7,145,370	Other cattle, . .	13,566,005
Mules and asses,	. 1,125,415	Sheep, . . .	28,477,951
Milch cows,	. . 8,935,332	Swine, . . .	25,134,569
Working oxen,	. 1,319,271		
	Total live stock, ...		. 85,703,913
	Valued at		. $1,525,276,457

The above chief products of the United States, raised in 1870, though they do not include all our products of agriculture, yet, large as they are, were produced on considerably less than one-half of the farm-lands of the country. Of the 407,735,041 acres occupied as farms, only 188,921,099 acres are improved, leaving 218,813,942 acres unimproved.

The capability of our soil to support the nearly forty millions of our present population and the millions yet to be born or to immigrate to this country, is very evident when we consider that less than one-half only of the farm-lands at present occupied were used to raise these immense products.

The rapid and general increase of products of the soil is very encouraging, and bespeaks for our country a prosperous future, if we but wisely use only a small part of the natural wealth of the soil and our mineral resources. From 1840 to 1870 the population increased 2¼ times, while the wheat crop increased 3.4 times, the corn crop more than doubled, and the oat crop increased 2.3 times. Thus our agricultural industries have more than kept pace with the increase of our population.

MANUFACTURES OF THE UNITED STATES.

Our manufactures are not less important than the products of agriculture, as will be seen by the following:

TABLE I.*

Of the Census Returns of Principal Manufactures of 1870:

Manufactures.	Establishments. Number.	Hands employed. Number.	Capital invested. Dollars.	Wages paid. Dollars.	Cost of materials. Dollars.	Value of products. Dollars.
Brass, cast, etc.,	319	4,703	6,793,435	2,411,232	5,252,262	10,459,735
Building (not marine),	24,908	112,820	49,168,863	49,741,376	95,694,685	201,572,541
Building materials,	33,207	251,582	233,005,203	73,126,900	173,198,451	356,140,945
Building (marine),	971	14,051	11,742,576	7,123,806	9,727,820	21,956,337
Copper milled, etc.,	99	1,991	5,506,800	1,123,558	13,201,289	15,796,750
Fire-arms,	709	5,835	9,583,502	3,667,202	5,140,181	12,748,419
Food and food preparations,	28,727	96,883	193,874,861	25,786,682	482,462,947	600,365,571
Furniture and house-fixtures,	6,312	57,091	46,766,983	23,304,956	28,516,544	75,539,719
Iron and its manufacture,	3,726	137,545	193,356,116	73,027,976	193,208,218	322,128,608
Lead,	91	843	4,822,100	408,903	13,703,921	18,327,196
Leather,	7,569	35,243	61,194,812	14,505,775	118,569,634	157,237,597
Lumber,	26,945	163,637	161,500,273	46,231,238	132,071,778	252,389,029
Paper,	684	13,779	35,780,514	7,477,780	31,344,169	50,842,445
Printing and publishing,	2,177	30,924	40,304,727	18,882,918	24,729,407	66,862,447
Textiles,	4,709	243,731	265,034,095	75,628,743	238,303,905	380,913,815
Articles of wear,	34,312	297,141	125,866,220	98,837,559	214,841,378	398,264,118
Liquors, spirits, malt, etc.,	3,089	19,060	66,658,945	9,009,062	49,110,288	94,123,014
Tobacco,	5,204	47,848	24,924,330	14,315,312	34,656,607	71,762,044
Total returns of man'factures U. S.	252,148	2,053,996	2,118,208,769	775,584,343	2,488,427,242	4,232,325,442

* Compiled from Compend. Census Repo t, 1870, ɪɴ. 873-884.

To appreciate the immense value of, and the progress made in, our manufactures, we have only to call
to mind that within the present century the people of
the whole of these United States numbered but little
more than are now engaged in manufacturing; that
one hundred years ago not a mill or a factory existed;
that the little manufacturing that was carried on by
our people was performed by hand, and was of the
coarsest kind; that the mother country was relied
upon for most of the necessaries, to say nothing of
the comforts, of life; that the whole region, now the
sites of most of our large cities and towns, was a
howling wilderness, inhabited by wild beasts, and by
men but little less savage.   It must also be recollected
that even in the civilized and older countries of
Europe all manufactures nearly a century ago were
produced by hand-labor.   Even England, with all her
present vast facilities and machinery for manufacturing, and the source of nearly all her wealth, is now
very nearly overtaken by the United States in the
race to become the world's workshop and factory.

TRADE AND COMMERCE.

The progress that has been made in the trade and
commerce of the United States during the last fifty
years has also been very great.

This can be seen by comparing the exports and imports of the several decades from 1810 to 1870, which
were as follows:

TABLE II.

Of Exports and Imports in years named:

Year.	Exports.	Imports.	Population.
1810, . .	$66,757,974	$85,400,000	7,239,881
1820, . .	69,691,669	74,450,000	9,633,822
*1870, . .	499,092,143	452,875,665	38,558,371

Fifty-four years ago (in 1820) the population of the United States was 9,633,822 ; in 1870 it was 38,558,371, or four times greater. In these fifty years our imports increased nearly six times, while the exports of our home productions are more than sixfold larger.

In twenty-eight years, from 1842 to 1870, the exports of the productions of the United States increased from $92,969,996 in 1842† to $499,092,143 in 1870. While the population is only little more than double, our exports of home productions have increased nearly fivefold.

Again, in 1842 the aggregate value of the exports of our domestic products was $92,969,996, while the value of like products exported in 1872 was $499,-092,143, or an increase in thirty years of more than fivefold in the exportation of the products of the United States to foreign countries. Again, while the exportation of our domestic products has increased, our importation of foreign products has comparatively decreased. In 1842 the value of our imports, exclusive of specie, was $96,075,071, or $3,105,075 more than the value of our domestic products ex-

* Commerce and Navigation Report, 1870, p. 615.
† Commerce and Navigation Monthly Reports, 1872, p. 132.

ported ; in 1872 the value of our imports was $640,·
337,540,* or $91,117,822 more than the value ($549,-
219,718) of the domestic products exported.

*   In 1872 among other domestic manufactures exported were the following :†

Cotton goods, .	. $2,304,330	Manufactures of iron, $6,812,383
Woollen goods,	. 212,669	" of steel, 1,934,723
Shoes and leather, .	. . . . . .	. 3,684,029

Thus the resources of our national wealth, of agriculture, manufactures, trade, and commerce, have
greatly increased.

The importance of fisheries in relation to the food-
supplies of a nation, and as a source of remunerative
employment, cannot be too highly estimated.

The great advantage arising from this source of
our nation's wealth is that the harvest of fish is
reaped without expense or preparatory labor ; the
fisher has only to gather and cure. To estimate the
importance of our fisheries as a food-supply, we have
but to state that in 1870, exclusive of our whale-
fisheries, their product was 1,135 bbls. sea-bass,
559,982 quintals of cod, 2,475 qtls. haddock, 10,955
qtls. hake, 2,451 tons halibut, 31,210 bbls. herring,
221,003 bbls. mackerel, 5,463 bbls. mullet, 647,312
bush. oysters, 3,216 bbls. pickerel, 24,118 bbls.
of salmon, 1,810,000 bls. canned salmon, 2,617
M. of shad, 69,561 bbls. of white-fish, 25,700 M.
white-fish, 766,930 gals. oil-fish, besides $1,208,778

---

* Commerce and Navigation Monthly Reports, 1872, p. 545.
† Ibid., pp. 546–48.

TABLE III.*—*Exhibiting the Mining Industries of the United States, 1870. This Table speaks for itself, and needs no comment.*

Minerals of States and Territories.	No. establ'ts.	No. hands.	Capital Invested.	Wages paid.	Materials, value.	Product.	Product, value.
			Dollars.	Dollars.	Dollars.	Tons.	Dollars.
Asphaltum,	1	23	514,286	20,000	26,773	(*)	450,000
Cinnabar,	4	811	11,900,000	599,000	30,700	(*)	817,700
Coal, anthracite,	231	53,096	51,016,785	23,015,813	3,600,540	15,604,275	38,495,745
Coal, bituminous,	1,335	41,658	58,991,244	21,300,678	2,063,415	17,199,415	35,029,247
Copper,	40	5,404	7,789,374	2,706,264	586,844	(*)	5,201,312
Gold, hydraulic min'd,	362	1,978	1,887,484	906,559	551,552	(*)	2,508,531
" placer mined,	1,632	8,463	5,624,549	2,472,020	1,306,782	(*)	7,266,613
" quartz,	224	3,297	9,454,500	2,185,862	446,209	(*)	4,360,121
" and silver quartz,	57	2,114	29,062,400	2,330,583	1,269,876	(*)	9,068,526
Iron ore,	420	15,022	17,773,935	6,838,022	1,279,563	3,895,718	13,204,138
Lead,	112	1,126	613,736	350,553	38,247	(*)	736,004
Marble,	22	795	1,316,600	442,270	36,916	(*)	804,300
Nickel,	1	48	60,000	6,400	1,800	(*)	24,000
Peat, cut,	4	39	13,100	5,550	550	(*)	8,200
Petroleum,	2,314	4,483	10,045,826	3,995,030	1,401,945	(†)181,263,505	19,304,224
Silver quartz,	102	1,056	4,015,000	841,501	489,914	(*)	3,248,861
Slate,	101	1,749	2,738,239	815,731	119,501	(*)	1,311,492
Stone,	997	12,573	7,152,854	5,322,183	979,124	(*)	9,971,100
Zinc,	15	588	2,414,942	250,075	40,440	(*)	788,880
Total, U. S.,	7,974	154,328	222,384,854	74,464,044	14,275,691		152,598,994

(*) Quantities not specified.   (†) Gallons.

TABLE IV.—*Statistics of the Fisheries of the United States, 1870.*

	No. establ'ts.	No. hands.	Capital Invested.	Wages.	Value of materials.	Value of product.
			Dollars.	Dollars.	Dollars.	Dollars.
All fisheries, except whale,	2,140	20,501	7,469,575	3,449,331	1,642,276	11,036,522

worth of miscellaneous fish. A visit to any of our great fish markets will aid any one to appreciate the food-wealth of our rivers and the sea.

This source of wealth in 1870 produced $11,096,522, and paid $3,449,331 for wages, and gave employment to not less than 20,504.

The art of fishing has been brought by degrees to its present perfect condition. In remote ages fish were caught by men who lay on the banks of rivers and on rocks, ready to shoot them with arrows or stick them with spears; even yet the partially-civilized take fish in blankets and sheep-skins.

RESOURCES FROM RAILWAYS, ETC.

The wealth of the nation invested in railways, and income derived from them, is immense, and is annually increasing. It is not more than fifty years since the building of the first railroad in the country was commenced. In 1830 there were only 23 miles of railroads in operation in the United States, but in 1873 there were 70,651 miles, whose gross earnings were $526,419,935.

TABLE V.[1]

The Miles of Railway in Operation, Amount of Stock, Receipts, Expenses, etc.

Divisions of the U. States.	No. miles R. Rs.	Total Capital Account.	Receipts from Passengers.	Receipts from Freight.	Operating Expenses.	Net Earnings.
		Dollars.	Dollars.	Dollars.	Dollars.	Dollars.
N. Engl'd,	5,314	263,697,778	22,358,645	29,318,043	36,614,911	15,061,777
Middle, .	14,019	1,126,702,107	42,355,230	151,697,072	124,771,717	69,280,585
Western, .	33,772	1,730,728,234	51,620,779	160,097,008	139,253,575	72,464,212
Southern,	15,353	509,324,106	15,456,162	38,456,162	35,551,060	18,145,349
Pacific, .	2,193	154,090,809	5,593,611	9,683,138	6,418,110	8,858,639
	70,651	3,784,543,034	137,384,427	389,251,423	342,609,373	183,810,562

* Poor's Railroad Manual for 1874-5, pp. 52-3.

It is estimated by Mr. Poor, from whose "Railroad Manual" the above data is taken, that at the rate of increase of the past three years the earnings of our railroads will be doubled in the next six years, without the construction of a single additional mile of road.

The population of the country is increasing about one million annually, and by the year 1880 the earnings of the roads now in operation will be not less than $1,000,000,000; the percentage of their earnings will be fully up to 20 per cent. of their cost. It is not likely that the construction of railroads will proceed as rapidly for a few years to come as it has done in the past; nevertheless, new lines will be constantly constructed, even in States that have the greatest length of lines in proportion to the population, to keep step with the rapid increase of population. Though millions of dollars have been lost in building railroads in advance of the wants of portions of the country or means for their support, the increase in the value of property due to their construction has far exceeded their cost. So that, if a few have lost, the nation has been the gainer.

The earnings of our railroads are but small when compared with the advantages they give the people in transporting persons and goods. The most distant cities have been brought near, and products that otherwise would be worthless have, by their construction, become chief contributors to our commerce and manufactures.

The following is, by the census of 1870, the estimated value of the personal property and real estate in the United States:

* The assessed value of real estate, . . .	$9,914,780,825
The assessed value of personal estate, . . .	4,264,205,907
Total assessed value of personal and real estate, .	$14,178,986,732
The true value of real and personal estate of the United States in 1870 was	$30,068,518,507

Allowing 25 per cent. for heedless and ignorant under-statement in 1860, when the true value of personal and real estate was stated to be $16,159,-616,068, to which add 25 per cent., or $4,039,904,017, for mistakes, it is gratifying to find that, even with this allowance, the true value of property, real and personal, in ten years has increased more than 50 per cent.

The annual income of the United States, by the census of 1870 and Poor's "Railway Manual" for 1873–4, is as follows:

From agricultural industries.	$2,447,538,658
" manufacturing industries, ...	4,232,325,442
" mining industries, ...	152,598,994
" fisheries, ...	11,096,522
" railroads in 1872,†	165,754,373
Total of these resources, ...	$7,009,313,989

The income of the manual-labor class, or the wages fund for 1870 of the United States, is as follows:

Wages—Agriculture during year, with value of board, ...	$310,286,285
" Mechanical and manufacturing industries,	775,584,343
" Mining industries, ...	74,464,044
" Fisheries, ...	3,449,331
▸ Total wages paid in 1870, by Census Report, .	$1,163,784,003

The receipts and expenditures of the Government for 1872 were:

*RECEIPTS.		EXPENDITURES.	
From Customs,	$216,370,286 77	Indians,	$7,061,723 82
Sale of Public Lands,	2,575,714 19	Pensions,	28,533,402 76
Internal Revenue,	130,642,177 72	Military Establishm'nts,	
Miscellaneous sour-		including Fortifica-	
ces,	15,106,051 23	tions, River and Har-	
		bor Improvem'ts,	35,372,157 20
To'l ord'y r'c'pts,	$364,694,229 91	Naval Est'blishm'ts,	21,249,809 99
Premium on sales		Miscellaneous, Civil,	
of coin,	9,412,637 65	includ'g Public Build-	
Receipts from loans		ings, Lighthouses, and	
and Tr's'y notes,	305,047,054 00	Coll'cting R'v'nue,	60,984,757 42
Gross receipts,	$679,153,921 56		
Balance in Treasury		Net ord'y expenses,	153,201,856 19
at the commence-		Premium on Bonds	
ment of the year,	138,019,122 15	purchased,	6,958,266 76
Cash previously re-		Interest on Public	
ported unavailable,		Debt,	117,357,839 72
since paid,	18,228 35	Public Debt,	405,007,307 54
		Gross expendi-	
		tures,	$682,525,270 21
		Bal'nce in Tr'sury at	
		the end of year,	134,666,001 85
Tot'l avail. cash,	$817,191,272 06	Total,	$817,191,272 06

A survey of the resources of the country, with its increase of population and wealth, must impress all with the pre-eminent advantages our people enjoy in developing its marvellous wealth and increasing the power and influence of the nation, and thus to make its people the most prosperous and happy on earth. The advantages which surround us for the procuring of wealth have never been enjoyed to so great a

* Report of Finances by W. A. Richardson, Secretary of Treasury of United States, for 1873, p. 12-17.

degree by any other people in the history of the world.

We should naturally expect that these immense resources would place all our people beyond the reach of want or destitution. But, alas! this is far from being the case; for though our nation's income is seven thousand million dollars ($7,009,313,989, the resources numerated above) per annum, yet we are compelled to witness a vast amount of destitution, crime, and pauperism visible in every part of the country. Why is it? Let us seek a solution.

CHAPTER II.

LABOR is man's inheritance, and his honor and glory. The world owes us nothing but what we labor for, though we all owe the world much.

At birth all are entitled to life, and no one has a right to interfere with it ; all are equally free born, and no one can justly subject us to his will.

Though we inherit a right to life, we have no right to live without producing, or to consume what is produced by others. The perfection of our organism brings with it at our birth needs that are more complex than other animals. Our food, clothing, and shelter must generally be extracted from the earth by labor ; and the greater our number of needs, the greater the necessity for labor.

The higher our civilization, the more numerous our wants, and the greater will be the labor required to supply them. To consume little, without producing what you consume, is to sink below humanity, and level ourselves with the lower animals.

To restrict ourselves to such food as merely supports existence, and labor only sufficient to obtain it, we rob society, and supply nothing toward repaying for the sacrifices made for us centuries before our birth. Without the labor of the past, the present race of human beings would be savages of a weak, perhaps servile, tribe ; for nearly all our present resources are the results of labor, nature having merely supplied us with the soil, air, water, and the spontaneous productions of the earth.

Man, under the most favorable natural circumstances, when left to battle alone for life, without the accumulated results of the labor of the past, is forced to endure a most wretched state of existence.

Nature has done a little for us, but labor much. Our simplest necessaries of the present, not to name the luxuries, have required the aggregate labor of centuries to produce. The race of animals now so serviceable to man, in aiding him in his labors, supplying him with food and clothing, once were wild, and roamed over the earth uncontrolled, except by animals more savage or in greater numbers.

The domesticated horse, which for unknown centuries has been in the service of man, once congregated in troops on the plains of Central Africa or Asia, from which the wolves and jaguars fled. The cow, sheep, and all the different breeds of domestic animals as now found, are the effects of man's labor and care in crossing.

The products of our fields, orchards, gardens, etc., are mainly the result of man's labor and perseverance. That most important plant, the potato, is not in its present condition a gift of nature, but by careful culture and labor man has brought it, step by step, to its present improved condition. Within the last century, and even before it was discovered in this country by the Spaniards, the Indians of Mexico and Peru had cultivated it. Wheat, the almost universal bread-corn, is not, as it now exists, the gift of nature, but is one of the master results of man's labor.

It has been generally supposed that Central Asia was the native country of our cultivated wheat ; but not many years ago M. Fabre, of Agde, South of France, discovered that the *Ægilops Ovata*, a grass of the regions near the Mediterranean and of the West of Asia, becomes, by cultivation, transformed into wheat. It grows spontaneously in Upper Egypt, but is there a poor, miserable seed, unfit for bread. It has, therefore, cost many, many centuries of labor to develop the bread we eat ; hence the nutriment of the wheat represents the blood, muscle, nerve-force, and life of thousands of generations of human beings who have perished during its culture. We hence owe a debt to those who have preceded us for the enjoyments by which we are surrounded. They all are the results of labor, and must be repaid in part by our labor, and by leaving the world better for our having existed.

The results of brain-labor have not been less useful than hand-labor, though they may not appear at first sight. Thinkers have done very much to extend the usefulness of hand-labor.

Though the hand of man has done much to bring him and his surroundings to the present exalted condition, yet his brain has done more. Thinkers, by their brain-work, have brought into existence the machines that lessen and facilitate labor.

The inventors of the jenney, the mule, and the power-loom did more useful work with their brains, and conferred greater benefits on mankind, than all the generations that spun and wove before they were born.

The discoverer of the process of making iron into steel, and Mr. Bessemer, who improved the process of its manufacture, rendered greater service to the human race during their lives than the most extensive manufacturers of steel.

The tools and implements used by man in all industries are an aggregation of ideas, the results of brain-work. Just in proportion as science and thought have developed, simplified, and improved the means of production, in that ratio has the brain-work of the past helped the present ; and the work of to-day, with all the accumulated results of thought, will help progress in the future.

Though the spinning-machines, power-looms, steam-engines, and all other machines were destroyed, the ideas that produced them still remain, and new ones could be made from the materials at hand. Hence it must be clear that the brain-work of the world is the most useful. The more our mental faculties are developed and cultivated, the more good we can do to others, and be the more useful to ourselves.

WEALTH AND LABOR.

The wealth of a nation is the aggregate result of the labor of the past. Wealth is anything that can be sold or will bring a price in the market, and represents a certain amount of labor that was expended to produce it. All wealth is the result of labor. Nothing possesses value until labor has been expended upon it. Raw materials are not exempt from this rule. Gold has no value while mixed with the sand in the bed of the river ; coal or iron is worthless and useless until labor has been expended upon it, and the labor expended is the measure of its true value.

The necessaries and conveniences of life annually produced and consumed are supplied by labor directly, or indirectly by purchase or exchange for some other product of labor.

A nation or people will be better or worse supplied with the necessaries and conveniences of life, according to what is produced and what is purchased, as they bear a greater or less proportion to the number of consumers. This proportion, in every nation, is regulated by two different circumstances :

1. By the knowledge, judgment, skill, dexterity, and the implements applied to the labor ; and,

2. By the proportion of those employed in useful labor to those not so employed.

Upon these circumstances will depend the abundance or scantiness of the annual products of a nation, whatever may be the extent of its territory, soil, or climate.

The measure of the gain or loss of an individual or a nation is the difference between what is produced and what is destroyed, either by economical consumption or extravagant waste.

All persons, whether productively or unproductively employed, as well as those who do not labor at all,
are maintained by the labor of the country.

When labor is properly applied, with the advantages of the accumulated discoveries of science and
the inventions in mechanic arts, the combined results
are marvellous.

The invention of labor-saving machines, and the results produced by them during the last century, have
been truly astonishing. The achievements of human
art in our own day have equalled those of Aladdin's
wonderful lamp. Though man is incapable of creating
or destroying a single particle of matter, yet the matter that composes the earth and the forces of nature
have been so utilized that they no longer resist human power, but are subdued to service. The steam-
engine alone, by the consumption of a ton of coal,
evolves as much force in one day as fifteen hundred
man-power. England, in 1854, from 50,000 tons of
coal derived the labor-force of two hundred and fifty
million able-bodied men ; hence the coal used in
Europe and the United States does nearly twice the
work that the whole population of the world could do
without it. Indeed, all calculations will fail to com-
pute the value of water-power and steam in conjunc-
tion with the numerous machines now used to multi-
ply human force. Nor is this all. Mechanical force is
matched by the increased motion that has been gained
by the fly-shuttles, hammers, rollers, wheels, etc. Mani-
fold rapidity is given in carding, spinning, weaving,
and printing, over the old one-thread spinning-wheel,

the hand-loom, and hand-press, by steam and machinery. Material forces, under the direction of machinery, have grown as light-limbed and as heavy-handed as man's needs have demanded.

Manufacturers, during the last century, have utilized the revelations of science and the inventions of machinery, with a corresponding increase in manufactured goods and a cheapening of productions, which has increased proportionately the happiness of the human family. This increased productive power has added to our general wealth, and our people are better provided with the commodities that supply life, the luxuries which refine it, and the masses are relieved from many severe drudgeries, thus giving opportunity for attention to higher mental and moral culture.

By a reasonable amount of labor and economy, with the accumulated powers of production, there is now no necessity for want of food or clothing. Though there has been, perhaps, less improvement in the machinery and implements of agriculture than in manufactures, yet the 5,922,471 persons engaged in agricultural pursuits in the United States are able to produce food for twice or three times our present population. One man can, by the aid of our farm implements, produce ample food for twenty persons. In the manufacture of clothing, with the extensive improvements that have been made in the application of machinery within the last hundred years, there is an almost incalculable productive power. If we take, for example, our cotton manufactures of 1870, we find that the 135,369 hands employed in cotton manufactures (of whom 22,942 were boys under 16, and girls under 15 years) produced 1,008,928,921 yards of the different kinds of cotton cloths. Allowing that an average of twenty yards are consumed annually by each man, woman, and child in the United States, the cotton goods manufactured in 1870 would be more than enough for fifty millions of people. By the same average, one person in a cotton-factory is able to produce cottons for the annual consumption of 372 persons. In the same year our woollen manufacturers employed 80,053 persons (9,643 of whom were under 16 years of age), who produced 188,588,688 yards of woollen cloths, besides millions of pounds of yarn and other woollen fabrics. To allow each male in the United States ten yards, the woollens produced in 1870 would be more than enough for all our male population. One person in a woollen factory can produce cloths for 235 males. This ratio will be applicable to the manufacture of all kinds of clothing, so that in the general production of clothing one person is able to supply all the wearing apparel needed by not less than fifty persons.

CHAPTER III.

In these days of inventions and the increased application of machinery it evidently should not be difficult to keep a person supplied with food and clothing. That men will not be satisfied to live upon the mere necessaries of life is readily admitted, neither is it needful that they should ; yet it is desirable that our citizens should so far appreciate their true interests and the advantages to health, wealth, and happiness, as to abstain from those expensive luxuries that merely gratify and create depraved and dangerous appetites.

In the early days of the settlement of this country the food was necessarily simple, clothing coarse, and the habitations rude ; indeed, this was comparatively the case in the older and more densely-populated countries of Europe. But as man progressed in knowledge, and as production was facilitated, all the necessaries of life have been increased in quantity and improved in quality ; so that everything really necessary for the happiness of mankind is now produced in great abundance. The only question of moment is their economical use and their proper distribution.

Though we have at command all things needed to secure the happiness of the whole human family, yet

the masses of mankind seem to be unhappy, and unknown numbers are suffering for the simplest necessaries of life. Why is this? All are interested in
the solution of this question ; for the sum of human
happiness will be incomplete so long as one member
of the human family is deprived of the necessaries or
comforts of life. In their eagerness to secure individual enjoyment and happiness men have generally
neglected the interests of their fellow-men ; and thus
have inevitably failed to obtain, in a great measure,
the happiness anticipated from the acquisition of
wealth, and the power and influence it would give.

The acts of the masses and governments have hitherto, in a greater or less degree, tended to produce
idleness, dissipation, and disease of both body and
mind, thus violating the natural laws and the principles of

POLITICAL ECONOMY.

Political economy, as a science, may be said to embrace the proper administration of the revenues of a
nation, the management and regulation of its resources, labor, productions, and property, and the
means by which the labor and the property of its citizens are protected and directed; as well as the best
methods of securing the success of each individual's
industry and enterprise, and general national prosperity.

Labor is generally divided by political economists
into two classes, viz., productive and unproductive.

Productive labor is that which adds directly to

value, as the labor of the shoemaker, mechanic, farmer, factory operative, etc., etc.

Unproductive labor is generally understood to be *labor* that is not employed in the production of wealth, or articles representing wealth, as soldiers, physicians, policemen, agents, school-teachers, etc., etc.

This classification, though made by most political economists, is not necessarily correct; for, under the present organization of society, the policeman and soldier, by adding security to wealth, stand somewhat in the place of producers. The laborer has no desire to work when he is not sure that he will enjoy the fruits of his labor ; and where wealth has not protection, it loses a great portion of its value. Hence the soldier and policeman add value to the products of labor by the security they give, and are not really unproductive. The teacher who labors to develop, to mould, to instruct the human mind is certainly not less a productive laborer than the blacksmith, stonemason, etc. And as man is of more value than iron or stone, no matter how much labor may have been expended upon them, so the labor of the educator is of more, much more, value than any amount of labor expended upon iron or stone.

It is also obvious that the physician who heals our diseases, and fits us again to return to some field of productive employment, is certainly not an unproductive laborer; for, without the aid of his skill, instead of being able to follow some productive employment, we should be destroyers, and not producers.

In reality, a non-producer is one who consumes without rendering an equivalent for what he destroys. All are destroyers, whether producers or not, for each must consume or destroy in order to prolong life ; and it depends upon what a man consumes in proportion to what he produces whether he is a productive or an unproductive laborer. For example, if a man consumes five hundred dollars' worth of products during the year, and has earned one thousand dollars, or produced products worth that amount, he is a producer and a productive laborer in the highest sense of the term ; for he is five hundred dollars richer. But if he spends five hundred dollars, and only earns that amount, he destroys as much as he produces, and is therefore a non-producer ; and if he should spend ten dollars a week, and only earn seven, he is not only a non-producer, but is a destroyer : he consumes more than he produces, and the country is three dollars a week the poorer.

All the wealth of a nation is not only the result of labor, but is also the fund out of which wages are paid for labor. The people of a country do not hoard their capital ; for those who have money generally use it for the gratification of their many wants, natural or acquired, and the wants of man usually increase with the power or opportunity for supplying them.

As labor is the source of a nation's wealth, prosperity, and power, it is obvious that whatever will give the most labor of the productive kind will be the best for the country. Whatever creates a demand for an article of productive industry also increases the demand for the labor to supply it, as well as a demand for labor to furnish the materials of which it is composed. For instance, the demand for broadcloth creates a demand for the wool, oil, dye-stuffs, and other articles used in its manufacture.

Again, before the cloth is consumed the labor of the tailor will be required, and also other materials to make it into garments. The influence of the demand for the cloth will be still further extended by the wages received by the tailor and the operatives by whom the cloth and the other materials were produced, which will be expended in supplying the wants of themselves and families, which again would give labor to other persons in various employments. Thus the increase of labor in one branch of productive industry will extend to and increase many other branches. That nation or people will be the most prosperous, other things being equal, who most encourage diversified productive labor, and discourage the use of, or the manufacture and the traffic in, whatever is useless or injurious.

The labor of the hatter is productive ; for he takes the wool or other materials of which the hat is made, and produces a useful and necessary article. Not only has his labor increased the value of the materials by making them into a hat, but by bringing the materials into use he has given them a new value. No matter how much labor it takes to produce an article, if it does not contribute to the health, comfort, or happiness of the consumer, it is unproductive. It is the general character of unproductive labor to consume the wealth of the nation without benefiting any one directly but the consumer, and rarely does even that. It is, therefore, of vital importance, in considering the question of labor, that the kind of labor should be kept in view ; for labor may be employed upon what will be useless and what will be injurious. Labor is a means and not an end. We labor for its beneficial results, for what it produces. Everything must be measured by its capability to administer to human comfort and happiness ; and any article that will not do this is labor lost in its production, and a waste of the materials composing it. Keeping these plain, common-sense views of wealth, labor, and consumption in our minds, we shall be prepared to enter understandingly into the merits of the subject before us.

CHAPTER IV.

THE QUANTITY AND COST OF INTOXICATING DRINKS.

AMONG the many evils resulting from the use of intoxicating drinks is the immense waste of money expended for them. The exact annual cost of these drinks in the United States can only be approximated, not ascertained.

The tax collected by the Internal Revenue Department in 1870 was upon 72,425,353 gallons of proof spirits (the specific gravity of 918.6, containing nearly equal weights of water and alcohol) and 6,081,520 barrels of fermented liquors.

CONSUMPTION OF LIQUORS.

Commissioner Delano, in his Internal Revenue Report for 1869, says: "In the absence of reliable data to fix the annual consumption of distilled spirits, we are left to conjecture. Were I to express an opinion on this subject, I should place the amount at not less than *eighty million gallons.*" This estimate of Commissioner Delano is corroborated by the Census Bureau, which reported ten years previous that there were produced in the United States, for the year ending June 1, 1860, 90,412,581 gallons of domestic spirits.

It is therefore safe to assume that the consumption of distilled spirits in the United States, in the form of beverages, is not less than the taxable quantity of spirits reported by the Internal Revenue Department in 1870, viz., 72,425,353 gallons. It may be said that a large portion of the annual production of spirits is used in the arts, exported, or used in various ways besides that of drinking. To this the following facts will furnish the answer:

1. We have taken about 8,000,000 gallons less than Commissioner Delano estimates to be the annual consumption.

2. The quantity taken does not include the large amount of liquors known to be made, and for which no tax is paid, as well as imported liquors that are smuggled into the country, which may safely be estimated at 5,000,000 gallons.

3. The liquors captured in the attempt to evade the payment of the tax, which, in 1870, amounted to 762,081 gallons of spirits and 10,310 barrels of fermented liquors.

4. The wines of California, the quantity of which is not given in official reports, but which the unofficial statements of the "trade" claim to be from 10,000,000 to 12,000,000 gallons.

5. The domestic wines made by farmers, which the Census Report of 1870 returns at 3,092,330 gallons.

6. Similar wines made from grapes, currants, and other fruits for private consumption, may be safely estimated, at least, at 1,000,000 gallons.

7. The difference between the above 72,425,353 gallons, which is proof-spirits, and the diluted or increased quantity, which, when dealt out to the drinker, is on an average not over 40 per cent. of alcohol, or

10 below proof, is equal to the addition of at least 7,500,000 gallons.

The quantity of the above liquors not embraced in official enumeration, amounting to 34,592,330 gallons, will clearly appear by the following tabulated statement, and will far exceed the quantity exported or used in the arts :

	Gallons.
Domestic spirits less than Commissioner Delano's estimate, ...	8,000,000
Domestic and imported liquors which evaded pay, .	5,000,000
Domestic wines, ...	10,000,000
Domestic wines made on farms, ...	3,092,330
Domestic wines made and used in private families, .	1,000,000
Dilutions of liquors, paying tax, by the dealers, .	7,500,000
Total, ...	34,592,330

It therefore may be safely said that in 1870 the liquors consumed, and their cost, were not less than as follows :

Domestic spirits, . .	72,425,353 gals.,* at 10 cts. a glass, or $6 a gal.,	$434,552,118
Ferm'nt'd liqu'rs, 6,081,-		
520 barrels, or . .	183,527,120 " " 5 " " " $20 a bbl.,	121,630,400
Imported spirits, . .	1,441,747 " " 10 a gal.,	14,417,470
" wines, . . .	9,088,894 " " 5 "	45,444,470
" spiritu's comp'nds,	34,239 " " 10 "	342,390
" ale, beer, etc., .	1,012,754 " " 3 "	3,038,262
	272,530,107	$619,425,110

* Dr. Young, Chief of the Bureau of Statistics, in a letter to Rev. Wm. M. Thayer, of Boston, said :

" In the absence of accurate data, the following is an estimate of the sales of liquors in the United States during the fiscal year ended June 30, 1871 :

Whiskey, ...	60,000,000 gals., at $6 retail,	$360,000,000
Imported spirits, ...	2,500,000 " " 10 "	25,000,000
Imported wine, ...	10,700,000 " " 5 "	53,500,000
Ale, beer, and porter,† . . .	6,500,000 bbls., " 20 "	130,000,000
Native brandies, wines, and cordials, unknown — estimated value of, ...		31,500,000
Total, ...		$600,000,000

"As a proof of the correctness of the above, it may be stated that during the

† By the Report of Internal Revenue for the fiscal year ending June 30, 1871, the tax was paid on 7,159,333 barrels of ale, beer, etc.

It must be clear that the above estimated cost of intoxicating beverages in the United States for 1870 is below the actual amount paid for them. This cost for drinks is nearly one-sixth of the value of the manufactures of the United States in that year, which was $4,232,325,442, and more than one-fourth of the value of all the "farm productions, betterments, and additions of stock," valued at $2,447,538,658. By the Census returns of 1870, the value of

Animals slaughtered and sold for slaughter was	$398,956,376
Home manufactures,	23,423,332
Forest productions,	36,808,277
Market-garden products,	20,719,229
Orchard products,	47,335,189
Total,	$527,242,403

Thus we find that the value of all the slaughtered animals, home manufactures, forest products, market-garden products, and orchard products was $92,182,707 less than the cost of our nation's drink-bill for the same period.

Again, by the Census returns of 1870, the value of

Articles of wear was	*$398,264,118
Furniture and house-fixtures (exclusive of stoves and hollow-ware),	†75,539,719
Total,	$473,803,837

which is $145,621,273 less than the cost of liquors for the same time.

last fiscal year the receipts from retail liquor-dealers, who paid $25 each for licenses, amounted to $3,650,000, indicating that there were 146,000 retailers of liquors in the United States. By including those who escaped paying license fees, estimated at 4,000, the number is increased to 150,000, who, on an average, sold at least $4,000 worth of liquors each, making $600,000,000, as above stated."

* Census Report, Vol. III., p. 485.   † Ibid., p. 437.

Thus in 1870 our NATION'S DRINK-BILL was one hundred and forty-six million dollars more than the estimated value at the place of manufacture of all the furniture and house-fixtures (except stoves and hollow-ware) ; all the boots and shoes, men's, women's, and children's clothing; all the collars, cuffs, gloves, mittens, hats, caps, hosiery, etc., etc., that were in that year manufactured in the United States.

Again, the value of all the food and food preparations of 1870 was $600,365,571, or $19,059,539 less in value at the place of manufacture than the cost of drinks.

If to the above value of food and food preparations be added 30 per cent. for profits of dealers, etc., before they reach the consumers, it will be $780,475,242. Then the food and food preparations consumed by the people of the United States in 1870 cost only $161,-050,132 more than the cost of the liquors drunk that year ; and if we include all the liquors consumed for which no tax or duty was collected it will be safe to say that more money is annually expended in the United States for intoxicating drinks than for all kinds of food consumed by the people.

Is it any wonder that tens of thousands of our people are in want of food and clothing, when there is expended annually for poisonous drinks as much as, or more than, is spent for food, and nearly twice as much as is spent for clothing ?

This needless waste is not for one year merely ; for our drink-bill increases and keeps pace with our population and productions, as will hereafter be more fully shown.

The quantity and cost of liquors that were entered for consumption and on market in 1871, as appears by the Internal Revenue Report and Report on Commerce and Navigation, were as follows:

	Proof-Gals.
Distilled spirits taken from bond, ...	59,503,972
Less spirits in market May 1, 1871, than in Nov. 15, 1870,	4,452,580
Total spirits entered for consumption, . . .	63,956,552
Domestic spirits exported, 1871, ...	971,313
Total spirits that paid revenue tax, . . .	62,985,239
Reduction of 20 per cent.* by dealers before sold, .	12,597,047
Total sold to consumers in 1871—gals., 40 per cent.,	75,582,286

	Barrels.
Domestic fermented liquors, ...	7,159,740
Domestic ale, beer, etc., exported, 1871, . . .	35,568
Leaving for home consumption, ...	7,124,172

	Proof-Gals.
Spirits and cordials imported, 1871, ...	2,478,845
" " exported, "	130,932
Leaving for home consumption, ...	2,347,913
Reductions of the above 20 per cent. before sold—gals., 40 per cent., ...	469,582
Leaving for home consumption—gals., 40 per cent. alcohol, ...	2,817,495

	Gallons.
Wines imported, 1871, †	10,422,904
Foreign wines exported, 1871, ...	138,252
Leaving foreign wines for home consumption, . .	10,284,652

* The liquors chemically examined by Prof. Draper of New York, in 1869, were found to contain on an average 40 per cent., and some as low as 22½ per cent., of alcohol; showing an average reduction from proof of 20 per cent., which would give an increase of 20 per cent. by dilution.

† Dr. Young estimates imported wines at 10,700,000 gallons.

The drink-bill of the United States for 1871 may be stated thus :

Domestic spirits, .   .   75,582,286 gals., at 10 cts. a glass, or $6 a gal., $453,193,716
Domestic ale, beer, etc.
(7,124,172 barrels),   . 213,725,160 "   "   5 "   "   "   20 a bbl.,   142,483,440
Imp't'd sp'ts & cordials,   2,817,495 "   " 10 "   "   "   10 a gal.,   28,174,950
Foreign wines,   .   . 10,284,652 "   at an average of $5 a gal.,   51,423,260
   "   ale, beer, etc., ... estimated,   4,460,676

Total liquors consumed,
   1871, ... 302,409,593 " costing consumers $680,036,042

The drink-bill for 1871 was $680,036,042, being an increase of $60,610,932 in one year.

THE QUANTITY OF LIQUORS AND THEIR COST IN 1872.

The reports of the Treasury Department for the fiscal year ending June 30, 1872, show that there were manufactured and imported into the United States alcoholic liquors as follows :

	Proof-Gals.
Domestic distilled spirits,*	69,033,533
Exported of the same,†	950,213
Balance, ...	68,083,320
Less spirits in market June 30, 1872, than in June 30, 1871, ...	1,512,516
Twenty per cent. of the above added for reductions by dealers, ...	13,616,664
Distilled spirits (domestic), reduced gals., 40 per cent. of alcohol, ...	83,212,500
Which cost the consumers, at 10 cents a glass, or $6 a gallon, retail, ...	$499,275,000

	Barrels.
Domestic fermented liquors,‡ . . .	8,009,969
Exported of the same,§	2,566
Leaving for home consumption, . . .	8,007,403

Costing the consumers, at 5 cts. a glass, or $20 a bbl., $160,148,060

* Internal Revenue Report, 1873.   † Commerce and Navigation Report, 1872.
‡ Ibid, 1872.   § Ibid.

Proof-Gals.

Spirits and cordials imported, . . . 2,131,837
Exported of the same, 1872, . . . 306,558

Balance, ... 1,825,279
Twenty per cent. added for the reduction
of alcohol, alcoh'c p'rc'nt'ge by dealers, 365,055

Leaving for home consumption, . . 2,190,334
at $10 a gallon, ... $21,903,340

Gallons.

Wines imported, 1872, * 9,863,313
Wine exported, 1872, ... 161,202

Leaving for home consumption, . . 9,702,111
Costing the consumers, at $5 a gallon at retail, . $48,510,555

Gallons.

Beer, ale, etc., imported, 1872, . . . 1,975,392
Exported of the same, 1872, . . . 14,361

Leaving for home consumption, . . 1,961,031
at $3 a gallon, ... $5,883,093

Our drink-bill for 1872 may be stated as follows:

Domestic spirits,	. .	83,212,500 gals.,	costing	$499,275,000
Domestic ale, beer, etc.				
(8,007,403),	.	240,222,090 "	"	160,148,060
Foreign sp'ts, cordials, etc.,	2,190,334 "	"		21,903,340
Foreign wines, .	. .	9,702,111 "	"	48,510,555
Imported ale, beer, etc.,	.	1,961,031 "	"	5,883,093
Total domestic and foreign liquors,	. .	337,288,066 "	"	$735,720,048

Thus in the year 1872 there were consumed in the United States 337,288,066 gallons of distilled spirits and fermented liquors, costing $735,720,048, being an increased consumption in one year of 34,878,473, and an increase in their cost of $55,684,006. .

Table VI., prepared with great care and much time

* Commerce and Navigation Report, 1872.

TABLE VI.*

States and Territories.	No. Distilleries, 1872.	No. Breweries operated, 1872.	No. Licensed retail liquor dealers.	No. Wholesale liquor dealers.	Aggregate sales of retail liquor-dealers at an average of $5,000 each annually.	No. Persons to each licensed liquor-dealer.	Male adults to each licensed liquor-dealer.	No. Gallons of distilled liquor produced, 1872.	No. Barrels of beer, etc., produced, 1872.
	(1)	(2)	(3)	(4)	(5)	(6)	(7)	(8)	(9)
Alabama.......	68	5	1,862	72	9,310.000	535	108	10,790	1,117
Arizona.......	......	10	241	40	1,205,000	41	15	....	512
Arkansas	22	1	1,582	70	7,910,000	306	64	40,432	125
California......	262	226	5,246	267	26,250,000	107	28	1,139,667	192,577
Colorado........	......	36	410	30	2,050,000	98	38	....	9,171
Connecticut....	55	25	3,778	143	18,890,000	143	34	360,172	57,416
Dakota.........	......	6	124	5	620,000	114	42	....	904
Delaware	13	2	326	4	1,630,000	383	87	5,419	4,268
Dist. Columbia.	......	13	1,097	54	5,485,000	121	29	....	15,056
Florida	......	2	694	19	3,470,000	270	56	....	124
Georgia	646	4	2,537	161	12,685,000	467	94	92,480	5,303
Idaho...	1	12	261	17	1,305,000	58	21	14,098	994
Illinois	98	216	8,918	302	44,590,000	285	61	19,471,852	497,917
Indiana........	121	169	5,061	146	25,305,000	332	74	7,043,866	158,957
Iowa..........	18	171	3,264	76	16,320.000	365	78	620,150	121,026
Kansas........	2	46	1,657	66	8,285,000	220	59	1,171	24,385
Kentucky......	237	46	4,446	389	22,230,000	297	64	5,257,101	101,404
Louisiana......	3	16	3,930	278	19,650,000	185	41	700,406	48,270
Maine..........	1	1	710	36	3,550,000	883	216	85,570	5,574
Maryland	24	72	4,629	289	23,145,000	169	37	1,556,201	173,531
Massachusetts.	23	56	10,031	477	50,155,000	146	32	2,840,755	570,432
Michigan	1	189	5,845	112	29,225,000	203	47	182,993	163,767
Minnesota.....	......	114	1,908	44	9,540,000	230	39	....	69,524
Mississippi.....	43	2	1,771	73	8,855,000	467	96	7,029	840
Missouri.......	91	124	5,922	313	29,610,000	291	64	2,287,285	368,968
Montana.......	......	36	357	33	1,785,000	58	32	....	2,567
Nebraska	......	23	586	21	2,930,000	210	62	209,032	16,568
Nevada	......	41	806	50	4,030,000	53	23	....	11,007
N. Hampshire..	3	5	915	39	4,575,000	347	91	40,800	101,310
New Jersey. ..	116	83	6,858	112	34,290,000	132	28	398,790	565,152
New York.....	92	479	26,744	1,156	133,720,000	164	36	4,766,154	2,602,505
New Mexico ...	2	8	397	40	1,985,000	234	57	223	737
North Carolina.	166	1	1,835	56	9,175,000	584	117	81,115	1 6
Ohio..........	110	233	11,401	452	57,005,000	234	52	14,708,029	725,609 1 0
Oregon	6	31	655	37	3,275,000	139	38	1,571	6,956
Pennsylvania..	86	443	15,745	841	78,725,000	223	50	2,231,004	1,006,828
Rhode Island..	1	4	1,010	55	5,100,000	213	43	64,276	17,808
South Carolina.	102	2	1,709	61	8,545,000	412	86	29,126	1,957
Tennessee	216	11	3,333	210	16,665,000	377	78	452,148	6,545
Texas..........	20	44	3,864	290	19,320,000	212	41	6,271	15,698
Utah	......	16	193	24	965,000	449	52	....	2,271
Vermont.	5	4	839	7	4,195,000	394	89	1,487	2,516
Virginia	312	13	2,856	141	14,280,000	425	93	432,563	10,562
Washington	......	14	292	20	1,460,000	82	27	13	4,130
West Virginia..	78	17	779	20	3,895,000	568	120	97,928	20,257
Wisconsin.....	10	292	3,607	109	18.035,000	292	56	1,019,330	295,818
Wyoming......	......	......	109	6	545,000	83	48	....	927
United States..	3,132	3,421	161,144	7,276	805,720,000	240	52	69,033,533	8,009,969

* In the preparation of this table the fractions were thrown out, except in a few cases where the fraction was large, when a unit was added.

and labor, is a compendious history of the liquor, traffic in the United States for the year ending June 30, 1872, based on the Internal Revenue Report for 1872 and the Census Returns for 1870. Column No. 1 shows the number of distilleries in operation; No. 2, the number of breweries; No. 3, number of licensed retail liquor-dealers; No. 4, number of licensed wholesale liquor-dealers; No. 5, the sales of licensed retailers of liquors, estimating $5,000 to be the annual average sales of each; No. 6, the number of persons for each licensed retailer of liquors; No. 7, number of male adults to each licensed retailer of liquors; No. 8, the gallons of spirits distilled; No. 9, the barrels of fermented liquors brewed during the year.

It will be seen by referring to the columns for Pennsylvania that there were in operation during the year 86 distilleries, 443 breweries; that there were 15,745 licensed retailers, 861 wholesale dealers; the sales of liquor $78,725,000, and one retailer to 223 persons, and one for every 50 adult males, etc. The same can be found for each of the States, etc.

QUANTITY AND COST OF LIQUORS FOR THIRTEEN YEARS IN THE UNITED STATES.

Having seen the approximated cost of liquors for the years 1870–71–72, we will now endeavor to ascertain our nation's liquor-bill for the 13 years between 1860 and 1872, inclusive. The following is an exhibit of the liquors reported to United States officials; also, an estimate of their cost, at the rates already given for the years 1870, 1871, and 1872:

Year.	Liquors paying Tax.	Retail Cost to Consumers.
1860*	. . . . 203,476,057 gallons, costing	$668,853,630
1861	. . . . †197,143,194 " "	613,433,995
1862	. . . . †191,954,182 " "	593,010,453
1863	. . . . ‡77,509,397 " "	181,593,317
1864	. . . . ‡209,554,922 " "	661,449,518
1865	. . . . ‡133,886,856 " "	208,996,925
1866	. . . . ‡181,391,444 " "	294,624,795
1867	. . . . §221,200,000 " "	600,000,000
1868	. . . . ‡172,117,445 " "	229,018,463
1869	. . . . ‡262,464,803 " "	693,999,509
1870	. . . . ‡272,530,107 " "	619,425,110
1871	. . . . ‡302,409,593 " "	680,036,042
1872	. . . . ‡337,288,066 " "	735,720,048

Total for 13 years, . 2,762,926,066 $6,780,161,805

The quantity of liquors reported for the 13 years
ending June 30, 1872, is much less than was consumed ; for not more than one-third of the liquors
manufactured in the United States during the years
1865–66–67 and 1868 were reported and paid duty to
the Government, as the examination of the following
table will fully establish :

TABLE VII.

*Showing the gallons of distilled spirits reported in the several years
by the Internal Revenue Department.*

Year.	Spirits.	Year.	Spirits.
1860	83,003,089 gallons‖	1867 ¶	14,575,168 gallons.
1861	No report of Inter. Rev.	1868 ¶	7,231,814 "
1862	" "	1869 ¶	62,092,417 "
1863 ¶	16,149,954 gallons.	1870 ¶	72,425,353 "
	(for 10 mos.)	1871 ¶	56,776,179 "
1864 ¶	85,295,391 gallons.	1872 ¶	69,033,533 "
1865 ¶	16,936,778 "	1873	71,151,367 "
1866 ¶	14,599,274 "	1874	69,572,062 "

* Census Report and returns of Custom-house for 1860.
† Domestic liquors estimated ; imported from returns of Custom-house for 1861.
‡ Reports of Internal Revenue and Commerce and Navigation for the years
given.
§ The estimate of Dr. Young, Chief of the Bureau of Statistics.
‖ From Census Report, 1860.   ¶ From Internal Revenue Report of that year.

In 1864, when the tax to March 7 was 20 cents, after that date 60 cents per gallon, 85,295,391 gallons were reported ; but when the tax was $2 a gallon, there were reported in 1865 only 16,936,778 gallons ; in 1866, 14,599,274 gallons ; in 1867, 14,575,168 gallons ; and in 1868, but 7,231,814 gallons.

In 1869 the tax was reduced to 50 cents per gallon, when there were reported 62,092,417 gallons, or 54,860,603 gallons more than in the previous year ; in 1870 there were reported 72,425,353 ; in 1871, 56,776,179 ; and in 1872, 69,033,533. Every one must feel certain that more liquors were manufactured and consumed than paid the tax of 1865–66–67 and 1868. It would be absurd to suppose that there was so great a falling off in the manufacture and consumption of spirits ; for in the very years when the greatest falling off of revenue occurred there was the greatest amount of drunkenness in our country. Every one knows that in 1865–66 and 1867 there was more intemperance than in any other years in the history of the country.

For in these years, just at the close of the war, the soldiers returned home with back pay, bounty, etc., much of which, if not the greater portion, was spent for drink, and went into the already well-filled tills of the drink-sellers. In regard to the falling off of the revenue on intoxicating drinks, there can be but one rational opinion, and that is that the Government was robbed of its just revenue by the liquor-manufacturers and liquor-dealers.

Dr. Young, Chief of the Bureau of Statistics, estimates the annual consumption (1867) to be about 221,200,000 gallons, and the cost about $600,000,000. He says : "These figures are sufficiently startling, and need no exaggeration. Six hundred million dollars ! The minds of few persons can comprehend this vast sum, which is worse than wasted every year. It would pay for 100,000,000 barrels of flour, averaging 2½ barrels of flour to every man, woman, and child in the country.

"This flour, if placed in wagons, ten barrels in each, would require 10,000,000 teams, which, allowing eight yards to each, would extend 45,455 miles— nearly twice round the earth, or half way to the moon. If the sum were in one-dollar notes, it would take one hundred persons one year to count them. If spread on the surface of the ground, so that no spaces should be left between the notes, the area covered would be 20,466 acres, forming a parallelogram of 6 by a little over 5½ miles, the walk around it being more than 22½ miles."

The truth, as the doctor says, will better serve the cause of temperance than any amount of exaggeration. The statements made by Dr. Young will greatly serve the cause of truth by enabling us to approach nearer to the true cost than we might otherwise dare to do. Dr. Young informs us that Mr. Wells's report of 1867, which gives $1,483,491,865 as the aggregate annual sales of the licensed retail liquor-dealers of the United States, includes other things sold by them as well as liquors. Though Mr. Wells's report does include other articles sold by retail liquor-dealers, yet it is nearer the real cost, direct and indirect, of liquors in the United States than any other official report yet made on the subject.

It is very probable that there are sold and consumed in the United States annually not less than 100,000,000 gallons of spirits. As already given, there were in 1860 88,003,089 gallons of spirits distilled. By the Internal Revenue Report of 1864 there were 85,000,000 gallons distilled during that year on which the tax was paid. No one believes all the liquor distilled is reported. Every report of the Internal Revenue mentions the seizures of illicit distilleries, etc., and the capture of thousands of gallons of liquors. In 1870 there were captured 762,081 gallons of spirits; and, notwithstanding the Brewers' Congress boasts of the great revenue they pay and their honesty, some of them attempted to rob the Government of the tax on 10,310 barrels of beer; and they too, with all their claimed honesty, like all other thieves, when they have "felt the halter draw" had no "good opinion of the law." No one supposes that all the frauds on the Government are detected, any more than all the thieves and pickpockets are caught in every act. Under the Revenue Laws of 1866–67, the distillers could and did systematically defraud the Government.*

Like an expert gambler, the distiller looks over the whole field and weighs the probabilities; the chances are nine out of ten that he will not be dis-

* See Report of the Select Committee of Congress on Internal Revenue Frauds, February 5, 1867.

turbed ; and if he should be so unfortunate as to be caught, he need not be alarmed.

Perhaps he has manufactured five or even ten thousand barrels, which he has disposed of without paying the tax ; he may have lost by the seizure fifty or sixty barrels, but he has put into his pocket perhaps a hundred and fifty or three hundred thousand dollars. This is a very strong inducement to run a small risk. When the case goes to Washington, he is promised, if he will pay the tax on the fifty barrels captured, with a small additional penalty, the proceedings will be stopped ; he does so, and the case is ended. The Committee on Internal Revenue Frauds reports that "among all seizures and prosecutions in the cities of New York, Philadelphia, and Brooklyn —and they have been many—your Committee cannot ascertain that a single case was pursued to the extreme limit provided by law."

The frauds upon the Government have not ceased, which makes it impossible for any one to ascertain the quantity of intoxicating drinks annually consumed ; all we can know is what is returned to the Internal Revenue Department.

A person may just as well say that there are only so many pickpockets because a certain number are caught, as that all the liquor made and consumed is reported and pays tax.

Therefore considerable allowance must be made when we attempt to estimate the extent of the liquor-traffic by official reports.

Of this we may be very certain : that the liquors

consumed are not less than the amount returned by the manufacturers and the dealers.

Dr. Young estimates the annual average sales of the licensed retail liquor-dealers to be about four thousand dollars ($4,000).

This average is certainly too small, for it is only $10 76 a day, which, if we allow them to make 100 per cent. profit, will only leave $5 38 per day. With this average daily income, every business man will readily see that the liquor-dealers would not be able to pay their high rents, license fees, taxes, the wages of bar-tenders, etc., support their families, and spend money freely, as they generally do. It is entirely out of the question for them to carry on their business as they do on an average of $5 38 per day profit. Any one who has given the subject much thought and investigation, or has any knowledge of the business, will not put their annual average sales at less than $6,000.

But, not to seem to over-estimate, we allow the average sales of the licensed retailers to be $5,000 per annum.

We venture to say, there is scarcely a three-cent liquor-den in Alaska or Baker Street, Philadelphia, or Five Points, New York, but will sell $5,000 worth of liquors during the year.

The average sales of the liquor-shops of the city of Philadelphia, if we leave out the unlicensed places, will not be less than $10,000 a year, which, with 4,105 licensed shops of 1867, will give as the annual liquor-bill of Philadelphia $41,050,000; and to take the licensed shops of 1873, which were 4,716, at the average of $5,000, then Philadelphia's liquor-bill is not less than $23,580,000 a year. Hence we feel confident that it will be no undue assumption for any community to average the sales of their licensed liquor-shops at $5,000 per annum.

This the liquor-men know to be true; if it is not, it is because the average is too low.

Before concluding this chapter, we ask to be excused for dwelling so long on this subject.

There have been so many different estimates of the cost of liquors that we have endeavored to furnish official and definite conclusions on the subject.

The above basis will furnish, we believe, a guide by which the cost of liquors in any community can be approximated sufficiently near for all practical purposes.

The standard average, $5,000 each, for the licensed drinking-places, considering the large number of unlicensed establishments which everywhere are the sure concomitants of licensed drinking-shops, will not give the cost of liquor more than it really is in almost any portion of the United States.

CHAPTER V.

IF we take Dr. Young's estimate (Chief of the Bureau of Statistics), that the annual cost of liquors in the United States is $600,000,000, then, for the ten years from 1861 to 1870 inclusive, our people spent six thousand million dollars ($6,000,000,000).

This drink-bill of six thousand million dollars for ten years—what an immense cost! What finite mind can grasp or comprehend the immensity of the labor-value of six thousand million dollars?

This sum appears as fabulous as the marvellous stories of the "Arabian Nights." But, alas! for poor humanity, for the welfare and prosperity of our people, and the honor of our country, it is no fiction, but a lamentable reality. This almost incomprehensible amount of money, produced by the sweat and toil of the toilers of our land, is spent mainly by our hard-working artisans, mechanics, and laborers, who can the least of all people in the country afford such prodigious waste and extravagance.

This hard-earned capital, that should be expended for food and clothing, for the half-starved and ill-clad thousands who are suffering for the want of them, or by the "parish bounty fed," is devoured by the demon of the still ; and, because this capital is so misspent, our jails are filled with criminals, our poor-houses with paupers, our asylums, homes, and charities with dependants, and our industrious, sober citizens burdened with taxes that would not be needed but for this waste of liquors. The cost of liquors for ten years is nearly two-thirds of the assessed value ($9,914,780,825) of all the real estate in the United States, while the assessed value ($4,264,205,907) of all the personal property of the United States is but little more than two-thirds of our ten years' drink-bill. Again, by the Census Returns of 1870, the value of all our "products of agriculture, betterments, and additions of stock ($2,447,538,658)," and the value of all our manufactures ($4,232,325,442), were the sum of $6,679,864,100, or only $679,864,100 more than is spent every ten years for liquors. Thus our people expend every eleven years for intoxicating drinks more than the value of all the products of agriculture and all our mechanical and manufacturing industries.

If in every eleventh year a fire should be kindled in the United States on the 1st of January, and continue burning until the last moment in December, and if every particle of our agricultural and manufactured products, as fast as they are produced, should be cast into the flames, and burned up until only the ashes remain, it would not inflict as much injury upon our people as is produced every eleven years by the use and sale of intoxicating drinks. The money expended for those drinks is not only lost, but the drinks entail upon our people the additional evils of vice, wretchedness, crime, and demoralization, that far, very far outweigh the value of the money expended for them. If the products to the value of the money spent for drinks were only destroyed by fire or flood, it would not deprive our industrious classes of the mental and physical power to replace them, as do the drinks for which their hard-earned millions are expended. What nation or people, however favored, can long exist and prosper who expend or waste the value of so much labor for poisonous drinks? Can we wonder that we have money-panics, hard times, and stagnation of trade?

The people who use such economy will ultimately become ruined and bankrupt. "The money," you say, "spent for liquor is not all taken out of the country, but is left to circulate among our people."

True. But what does the purchaser receive for his money spent for drinks? Absolutely nothing. *Ay, worse than nothing;* for they do not promote his health, comfort, or happiness, but injure his health, mar his comfort, destroy his happiness, unfit him for productive labor, shorten his life, and militate against all his interests, for time and for eternity.

Again : The capital spent for alcoholic drinks adds nothing to the consumer's possessions, as do wholesome food, clothing, furniture, and other property, real and personal ; it is spent for poisonous slops, that give but momentary excitement to his animal passions or sentient pleasures, and finally leave him physically, mentally, and morally worse for their use,

And it would have been vastly better for him if he had cast the money into the fire, or had poured the liquor into the gutter as soon as he had paid for it. He would have felt no loss, but would be a gainer by so doing.

THE NATION'S LOSS BY THE DRINK-TRAFFIC.

The loss to the nation by the use of intoxicating drinks and the traffic in them is incalculable, but is certainly not less than the money paid for the drinks, and we should be no worse off if we should suspend our national industries to the value of the money paid for liquors, providing that at the same time we entirely ceased their manufacture and sale.

To illustrate this, let us compare the industries of a few of the States with the cost of liquors to the people of the same States :

States.	Value of the Products of Agriculture.	Value of Manufact'r's.	Total Wages Paid.	Cost of Liquors to Consumers.	Receipts of the Railways in 1873.
	Dollars.*	Dollars.*	Dollars.*	Dollars.*	Dollars.*
New York, .	253,526,153	785,194,651	176,918,120	106,590,000	68,825,007
Pennsylvania,	183,946,027	711,894,344	151,158,538	65,075,000	83,357,427
Illinois, . .	210,860,585	205,620,672	53,439,011	42,825,000	54,086,412
Ohio, . .	198,256,907	269,713,610	65,547,260	53,845,000	59,508,950
Massachusetts,	32,192,378	553,912,568	123,872,918	25,195,000	27,850,458
Maine, . .	33,470,044	79,497,521	17,185,497	4,215,000	4,363,741

By examining the above figures, given in the Census of 1870, Table 86, etc., it will be seen that in the year 1870 there was spent for liquors in New York $106,-590,000, or more than two-fifths of the value of products of agriculture, and nearly one-seventh of all

* See Table VIII., from which they are taken.

TABLE

Exhibits Number of Licensed Retail Liquor-Dealers, and the Estimated Agricultural Products and Wages; also, the Manufactures, with Products in each State and Territory, by the Census of the

STATES AND TERRITORIES.	Number of licensed retail liquor-dealers.	The estimated cost of liquor to the consumers; estimating that the average sales of each licensed retail liquor-dealer are $5,000 per annum.	Agriculture of United States. Production and Wages.*	
			Estimated value of all farm productions, betterments, and additions to stock, by the Census of United States, 1870.	Total amount of wages paid, including the value of board.
	Number.	Dollars.	Dollars.	Dollars.
1 Alabama............	1,976	9,880,000	67,522,335	11,851,870
2 Arizona............	119	595,000	277,998	104,620
3 Arkansas..........	2,000	10,000,000	40,701,699	4,061,952
4 California.........	5,845	29,225,000	49,856.024	10,369,247
5 Colorado.....	371	1,855,000	2,385,106	416,236
6 Connecticut.......	3,352	16,760.000	26,482,150	4,405,064
7 Dakota............	82	410,000	495,657	71,156
8 Delaware.........	3,8	1,840,000	8,171,667	1,696,571
9 District Columbia..	1,087	5,435,000	319,517	124,338
10 Florida	580	2,900,000	8,909,746	1,537,060
11 Georgia...........	2,767	13,835,000	80,390,228	19,787,086
12 Idaho.............	244	1,220.000	637,797	153,007
13 Illinois	8,562	42,825,000	210,840,585	22,338,767
14 Indiana...........	4,444	22,220,000	122,914,302	9,675,348
15 Iowa	3,073	15,365,000	114,386,441	9,377,878
16 Kansas...........	1,117	5,585,000	27,630,651	2,519,452
17 Kentucky	4,761	23,805,000	87,477,374	10,709,382
18 Louisiana........	4,414	22,070,000	52,006,022	11,042,789
19 Maine.............	843	4,215,000	33,470,044	2,903,292
20 Maryland	4,285	21,425,000	35,348,927	8,560,367
21 Massachusetts	5,039	25,195,000	32,192,378	5,821,032
22 Michigan	5,020	25,100,000	81,508,623	8,421,161
23 Minnesota........	1,930	9,650,000	33,446,400	4,459,201
24 Mississippi.......	1,807	9,035,000	73,137,953	10,326,794
25 Missouri..........	5,888	29,440,000	103,035,759	8,797,487
26 Montana..........	449	2,445,000	1,676,660	325,213
27 Nebraska.........	635	3,175,000	8,604,742	882,478
28 Nevada...........	658	3,290,000	1,659,713	438,350
29 New Hampshire ...	1,161	5,805,000	22,473,547	2,319,164
30 New Jersey.......	5,649	28,245,000	42,725,198	8,314,548
31 New Mexico.......	418	2,090,000	1,905,060	523,828
32 New York.........	21,318	106,590,000	253,526,153	34,451,362
33 North Carolina....	1,315	6,575,000	57,845,940	8,342,856
34 Ohio..............	11,769	58,845,000	198,256,907	16,460,778
35 Oregon	738	3,690,000	7,122,790	719,875
36 Pennsylvania......	13,015	65,075,000	183,946,027	23,181,944
37 Rhode Island......	727	3,635,000	4,761,163	1,124,118
38 South Carolina....	1,565	7,825,000	41,999,402	7,404,297
39 Tennessee	2,684	13,420,000	86,472,847	7,118,003
40 Texas............	2,168	10,840,000	49,185,170	4,777,638
41 Utah	128	640,000	1,973,142	133,695
42 Vermont...........	540	2,700,000	34,647,027	4,155,385
43 Virginia..........	3,314	16,510,000	51,774,801	9,753,041
44 Washington.......	224	1,120,000	2,111,902	215,522
45 West Virginia......	543	2,715,000	23,379,692	1,903,788
46 Wisconsin....	3,864	19,320,000	78,027,032	8,186,110
47 Wyoming.........	236	1,180,000	42,760	3,075
United States....	143,115	715,575,000	2,447,538,658	310,286,285

* Census Report, Vol. III., p. 81,

VIII.

Cost of Liquors in each State and Territory ; also, the Value of all the Capital invested, Wages paid, Cost of Materials, and Value of the United States, 1870.

	Manufactures of the United States.*				
	Wages paid.	Capital Invested.	Cost of material.	Value of Production.	The total receipts of the railways of the U. S. for 1873, from Poor's Railway Manual of the U. S. for 1874-75.
	Dollars.	Dollars.	Dollars.	Dollars.	Dollars.
1	2,227,968	5,714,032	7,592,837	13,040,644	4,957,941
2	45,580	150,700	110,090	185,410	....
3	673,963	1.782,913	2,536,998	4,639,234	927,609
4	13,136,722	39,728,202	35,351,193	66,594,556	15,276,749
5	528,221	2,835.605	1.593,280	2,852,820	1,098.596
6	38,937,147	95,281,278	86,419,579	161,065,474	10,544,810
7	21,106	79,200	105.997	178.570	162,725
8	3,692,195	10,839 093	10,206,897	16,791,382	666,801
9	2,007,600	5,021 925	4,754,883	9,292,173	....
10	949,592	1 679,980	2,330,873	4,685,403	479,000
11	4,844,508	13,930.125	18,583,731	31,196,115	7,695,955
12	112.372	742 300	691,785	1,047,624	....
13	31,100,344	94,368,057	127,600,077	205,620,672	54,086,412
14	18,366 780	52,052,425	63,135,492	108,617,278	24,279,062
15	6,893 292	22 420,183	27,682,096	46,534,322	7,983,988
16	2,377,511	4,319,060	6,112,163	11,775,833	10,062,437
17	9,444,524	20,277,809	29,497,535	54,625,809	7,199,993
18	4,593,470	18,313,974	12,412,023	24,161,905	2,740,489
19	14,282,205	39.796,190	49,379,757	79,497,521	4,363,741
20	12 682,817	36 438,729	46,897,032	76,593,613	15,310,942
21	114,051,886	231.677,862	334,413,982	553,912,568	27,850,458
22	21,205,355	71.712,283	68,142,515	118,394,676	14,295,088
23	4.052,837	11,993.729	13,842,902	23,110,700	4,212,844
24	1,547,428	4,501,714	4,364,206	8,154,758	5,424,326
25	31,055,445	80,257,244	115,583,269	206,213,429	12,189,908
26	870 848	1,791,300	1,316,331	2,494,511	....
27	1,429,913	2,169,963	2,902,074	5,738,512	11,358,447
28	2,498,473	5,127,790	10,315,984	15,870,539	....
29	13,823,091	36,023,743	44,577,967	71,038,249	8,618,460
30	32,648,409	79,606,719	103,415,245	169,237,732	25,840,923
31	167,281	1,450,695	880,957	1,489,868	....
32	142,466,758	366,994,320	452,065 452	785,194,651	68,825,007
33	2,195,711	8.140,473	12,824,698	19,021,327	2,897,488
34	49,056,488	141,923,964	157,131,697	269,713 610	59,508,950
35	1,120,173	4,376,849	3,419,756	6,877,387	....
36	127,976,594	406,821,845	421,197,673	711,894,344	83,357,427
37	19,351,256	66,557,322	73,154,190	111,418,354	1,115,672
38	1,513,715	5,400,418	5,355,736	9,858,981	3,560,027
39	5,390,630	15,595 295	19,657,027	31,362,636	4,451,517
40	1,797,845	5,284,110	6,273,193	11,517,302	6.147,648
41	895,365	1,391,898	1,238,252	2.343,019	1 832,612
42	6,264,581	20,329,637	17,007,769	32,184,606	4,183,547
43	5,343,009	18,455.400	23,832,381	38,364,322	7,098,243
44	574,936	1,893,674	1,435,123	2,851,052	....
45	4,322,164	11,094,520	14,503,701	24,102,201	51,202
46	13,575,642	41,981,872	45,851,266	77,214,326	11,146,812
47	347,578	889,400	280,156	765,424	....
	775,581,343	2,118,208,760	2,488,427,242	4,232,325,412	526,419,935

* The same, p. 392.

manufactures, and nearly two-thirds of the wages paid for both agriculture and manufactures; the liquor-bill being little less than twice the receipts of her railroads.

The liquor-bill of Pennsylvania in 1870 was $65,-075,000, which was one-third the value of the products of agriculture, nearly one-tenth of her manufactures, more than two-fifths of wages paid, and about three-fourths of the receipts of her railroads, though there are more miles of railroads in Pennsylvania than in any other State in the Union.

The liquor-bill of Illinois was $42,825,000, or more than one-fifth the value of the products of agriculture, a little less than one-fourth of her manufactures, about ten million dollars less than the aggregate wages paid for all the agricultural and manufacturing industries of the State, and only about eleven million dollars less than the annual receipts of her railroads. That year Ohio paid for liquors $58,845,000, which was more than one-fourth of the value of the products of agriculture, and more than one-fifth the value of her manufactures; while it was only a little more than six million dollars less than all the wages paid for labor, and as much nearly as the receipts of all the railroads of the State.

The liquor-bill of Massachusetts was $25,195,000, being five-sixths of the value of her products of agriculture, one-twenty-second of the manufactures, and not one-fifth of the amount of wages paid.

In Maine liquors cost only $4,215,000, or less than one-eighth of the value of the products of agriculture, less than one-sixteenth the value of the manufactures, and only two-sevenths of the wages paid in the State.

It is said by the friends and supporters of the liquor-traffic that the prohibitory laws of Maine and Massachusetts are failures. Now, we ask these persons to examine the above carefully, and compare the liquor-bills of these two States with the products of agriculture, manufactures, and wages paid ; and then turn to the other six States given, or to any State granting license, and make the same comparisons, and the result will in each case be found in favor of the State which most restricts the liquor-trade. If all the States and Territories be so examined—which can be easily done by consulting the preceding Table VIII., which exhibits the amount of the cost of liquor to each State, and then compare the cost of liquors with the value of the products of agriculture, manufactures, and wages paid in each—the conclusion will be inevitable that the true policy of any State seeking the prosperity and happiness of its people is prohibition.

Can any reflecting person consider the immense cost of alcoholic drinks, when compared with the value of the productive industries of the nation, without forebodings of the ruin that is in store for our country ? This waste does not exist alone in Pennsylvania or New York, but the same cause is producing the same effects in all States of the Union, as can be seen by Table VIII.

Is it possible for any state or nation to long prosper

or exist whose people spend for demoralizing drinks so large a proportion of the value of their products of industry? The people who practise such irrational *political economy* will eventually sink into decay, and leave a mass of mouldering ruins as monuments of the egregious folly of allowing or licensing a traffic that produces as legitimate fruits idleness, poverty, crime, disease, and death.

COST OF WAR AND DRINKS.

From an essay furnished by David A. Wells to the Cobden Club, England, upon the expenses, income, and taxes of the United States, we learn that the whole cost of the war of the Rebellion, North and South, from 1861 to 1866, is estimated as follows: Lives, 1,000,000; property by destruction, waste, etc., $9,000,000,000. The expenditures of the United States from June, 1861, to July, 1866, $5,792,257,000; of this the actual war expenses were about $5,342,237,000. The expenses of States, counties, cities, and towns in the Northern States, not represented by funded debt, have been estimated at $500,000,000. The increase of State debts on the war account was $123,-000,000. The increase of city, town, and county debts is estimated at $200,000,000; the total war expenses of the loyal States and National Government, $6,165,237,000. The estimated direct expenses of the Confederate States on account of the war were $2,000,000,000. Aggregated expenses of the country, North and South, $8,195,237,000. The total receipts from all sources during the second year of the war were less than $42,000,000. The expenditures were $60,000,000 per month, at the rate of $700,000,000 per annum.

This immense cost of treasure and blood during the five years of the Rebellion is truly appalling. Yet it was not all spent in vain, for the nation was saved and chattel slavery abolished.

But the slavery of strong drink rules our country yet. Its slavery is vastly more oppressive, more degrading to its victims, and much, very much, more injurious to their moral, religious, and intellectual condition and the general financial affairs of the country, than the chattel slavery of the Southern States. But let us take a glimpse at our nation's drink-bill compared with our late war expenses.

The annual cost of intoxicating drinks in the United States, at Dr. E. Young's estimate of $600,000,000 a year, in ten years would amount to the total war expenses of the loyal States and the National Government. Our drink-bill in thirteen and a half years would amount to more than the aggregate war expenses of both the North and the South.

Every fifteen years we expend more for strong drinks than the value of all the property wasted and destroyed during the five years of the war. And every year it costs our people over one million dollars more for strong drinks than the expenses during the war of all the States, counties, cities, and towns in the Northern States not represented by funded debts.

When we were expending $60,000,000 a month, or

at the rate of $700,000,000 a year, to crush out the Rebellion and save the nation, a cry arose in all parts of the country that the nation would be bankrupt, that we should never be able to pay off the debt. Men all over the land, on every public and private occasion, at every gathering, railed at and found fault with the Government because of these expenditures. Yet, strange as it may seem, we were expending at the same time almost as much, if not more, for poisonous beverages or strong drinks, without a single word of complaint. It is very safe to say that in every year from 1861 to 1874 more money was spent for alcoholic drinks than would pay the annual war expenses.

Again, the money now spent for strong drinks, if devoted to the liquidation of our national debt, would pay it all off in less than three years.

What a deplorable cost! What a shame that a professedly Christian nation should pay annually six or seven hundred million dollars to produce poverty, crime, degradation, demoralization, financial depression, and ruin !

No people were ever so favored by a beneficent Creator.

What do our people receive of these millions expended for liquor? First, they receive about 274,-456,376 gallons of a great variety of admixtures of various degrees of alcoholic strength. The liquor drunk in 1870 would allow about seven gallons for each man, woman, and child of the Union.

Why do our people drink it? It cannot certainly

be for the watery portion, for much better and purer water can be obtained freely from the rill and the spring.

If it is not for water, what are its other ingredients? It contains, besides spoiled water, alcohol, which is an acrid narcotic poison, and frequently numerous other drugs. But it is drunk for the excitant alcohol. How much alcohol do our people drink? If distilled spirits contain 50 per cent. of alcohol, malt liquors 5 per cent., and wines 20 per cent., our people will consume of pure alcohol, contained in the domestic and imported spirits, ale, beer, wine, etc., the quantity of which has heretofore been given, as follows, viz. :

	Gallons.
In distilled spirits, cordials, etc., . . .	38,798,202
In malt liquors, ...	9,948,310
In wines, ...	1,908,666
Total gallons of alcohol consumed, . .	50,655,178

Thus there are consumed in our country annually more than five quarts of the purest alcohol for every man, woman, and child. Can we wonder at the large sickness and death-rate in our country when so much liquid poison is drunk? We are confident that satisfactory evidence can be adduced to show that *but for the use of intoxicating drinks the sick-rate, as well as the death-rate, of our country would be reduced to not less than one-half what it is at present.* If all the alcohol were taken out of these drinks, what would be left for the money paid for them?

There would remain 227,874,935 gallons of spoiled
water, or, more strictly speaking, a decoction of vile,
poisonous drugs that the very swine would refuse to
drink, unless their appetites and tastes had first
become perverted, by being starved in a brewery or
distillery, until they learned to drink.

What else have we for all the millions spent for
them? We have drunkenness, with its follies, its
revels, its obscenity, its beastliness, crime, and taxa-
tion. These drinks are marshalled enemies against
civilization, liberty, justice, humanity, morality, and
religion. In addition to all these we have also
poverty, with its attendant evils; ignorance, with its
vulgarity, brutishness, and vices; crime of every
degree ; accidents on land and sea; idiocy, insanity,
madness, disease, premature old age, and death. Are
all these worth the price paid for them? Does it pay?
In the prayerful hope that the statistics and their
relations and comparisons with the drink-traffic may
prove the looking-glass by which the blotched and
pale face of the body politic, weakened by intoxicating drinks, may clearly be seen, the author has
toiled many weary hours in presenting them to his
countrymen.

CHAPTER VI.

The drink-bill of the United States in 1870, exclusive of domestic wines and other liquors made and drunk, was $619,425,098 ; and, in addition to this sum, it must be clear to every one who knows anything of the drink-trade that more liquors are annually manufactured and sold than are reported to the Government.* In consideration of the difficulty of being able to arrive at the true cost of intoxicating drinks, we have estimated the annual sales of licensed retail liquor-dealers at $5,000 each, and it is assumed this average will furnish a correct basis for ascertaining the annual cost of liquors in the United States.

There were in 1870 143,115 licensed retailers, whose aggregate sales of liquors on the above basis would be $715,575,000.

The difference between this sum and that of the estimate based on the returns of the Internal Revenue Report of 1870 is $96,149,902, which will be a small allowance for the liquors which are made and consumed, but not embraced in any official report.

This $715,575,000, taken from the productive industry of the country, is in itself sufficient to cause

---

* The liquor frauds just brought to light all over the country plainly prove the impossibility of arriving at the true cost of intoxicating drinks from Revenue Reports. At best we can but approximate ; hence all our estimates are below the true cost.

71

great depression of our national trade. All that is needed to secure the most abundant commercial prosperity, and to give full employment to all classes of productive laborers, is to transfer these millions from the liquor business, and devote them to the purchase of the necessaries of life, which would increase the demand for the products of agriculture and manufactures equal to the value of the sum thus uselessly expended.

If our cotton, woollen, and other factories are not now as busy as formerly, it is because they cannot dispose of their goods. It is also certain that if the goods are not disposed of, it is because the people either spend their money for other articles, or are poorer and have not the money to spend. Our people cannot be poorer; for year after year the wealth of the country increases faster proportionately than does the population. Wages are higher in this than in any other country.

The truth of the matter is that our people squander their money for what is not only useless, but injurious, and devote their wages to the purchase of those things which give the least labor to our workers.

Let us examine just one item of lavish and improvident expenditure—the use of intoxicating drinks—and see if this does not chiefly cause the bad trade and the severe depression of business now so general all over our country.

During the years 1869–70–71–72 there was expended in the United States for intoxicating drinks as follows:

In 1869,	.	.	.	.	.	.	$693,999,509
In 1870,	.	.	.	.	.	.	619,425,110
In 1871,	.	.	.	.	.	.	680,036,042
In 1872,	.	.	.	.	.	.	735,726,048
Total for 4 years,	.	.	.	.	$2,729,186,709
Annual average,	.	.	.	.	682,296,677

Being an average for those four years of more than six hundred and eighty-two million dollars spent annually for intoxicating drinks.

It is self-evident that he who spends his wages for drink, unless he is richer than the majority of our citizens, must be deprived of many luxuries, and even necessaries, that he could have procured for himself and his family if he had not so spent his money. Therefore, while drinking-shops flourish and increase, the business in every department of productive industry must languish and decline.

There is an English proverb which says: "A fool can make money; but it requires a wise man to spend it." The people of our country, as well as almost every other country, like the fools in the proverb, can make money, but do not spend it like wise men. For we find, in 1870, that the wages paid for labor in all our manufacturing industries were $775,-584,343 ; and that in the same year there were spent not less than $619,425,110 for strong drinks. Was this money spent wisely? Our population in 1870 was 38,558,371, and the value of some of our manufactures was : * Textiles, $380,913,815 ; articles of wear, $398,264,118 ; boots and shoes, $181,644,090 ;

* See Vol. III. of Census Report, 1870—"Wealth and Industries."

cotton goods, $168,457,353; woollen goods, $151,-298,196.

If all these commodities produced in 1870 were consumed in that year—which is very likely—or at least their value in like products, domestic or foreign; if we add 25 per cent. to the values above given for profits and expenses before they reach the consumers, then for each man, woman, and child in the United States there were spent for our textiles—which includes cotton goods, flax and linen goods, carpets, woollen goods, and worsted goods—$12 30, and for liquors about $16 06; and for each family of 5.09 persons* there were expended for textiles $62 60, and for strong drink $81 74. For articles of wear—including men's, women's, and children's clothing, boots, shoes, hats, caps, collars, cuffs, gloves, mittens, hoop-skirts, corsets, and hosiery—there were spent for each person $12 91, for liquors $16 06; for these articles of wear for each family were expended $65 71, and for liquors $81 74. There were expended for boots and shoes for each person in the United States $5 89; for liquors, $16 06; and for each family for boots and shoes, $29 98; for liquors, $81 74. For cotton goods of every description there were expended for each person $5 46, and for each family $27 70. For woollen goods for each person, $3 23; and for each family, $16 44. While, as already seen, for liquor there were spent for each individual $16 06, and for each family $81 74.

---

* The average of the families in United States is 5.09 persons.

Is not this wasteful expenditure for a pernicious article clearly the cause of bad trade and business panics? It is very certain that our people cannot pour upwards of seven hundred million dollars annually down their throats in the shape of alcoholic drinks, and spend them also for clothing. "You cannot eat your cake and have it also," says the proverb. Nor can a man encourage productive industry who spends his money for drink ; for every cent left with the rumsellers is taken from the butcher, baker, shoemaker, tailor, etc., and is lost to productive labor.

The value of the food and food-preparations of 1870 was $600,365,571,* or $15 57 for each person, and $79 25 for each family. Thus our people spend nearly as much for liquors as for foods ; for we have already seen that $81 74 are spent by each family for liquor.

It is always the effect of non-productive labor to waste capital, whilst productive creates capital. Capital furnishes food, raiment, shelter, etc., for the laborer and his family, besides providing implements and machinery to aid in his labor, and the raw materials for the articles he produces. Whatever aids to produce these necessaries, now or in the future, is capital. Without capital there can be no productive labor ; no combination or division of labor ; nor those implements of industry and machinery that enable us to overcome the many obstacles which have stood and still stand in the way of genius and industry.

It is also an indispensable prerequisite for the promotion of physical, intellectual, moral, and religious culture and progress.

* Without the profits of dealers, etc.

It must be clear to every reflecting person that whatever causes an unproductive expenditure of capital will retard progress. This is eminently the case with money spent for liquors.

Intoxicating drinks are not only unnecessary, but the money spent for them is so much capital taken from those branches of industry that add to the growth and the prosperity of the nation.

The money spent for these drinks, if expended for useful articles, would afford increased employment for laborers in every department of life. If it were not for the drinking customs in our country, there would be no lack of work at remunerative wages for all our unemployed, and even those now engaged in the manufacture and sale of intoxicating drinks could be engaged in better and more honorable kinds of business.

It is said that a man once applied for work to a Philadelphia millionaire, who, not having a situation for him just then, but liking his appearance and wishing to help him, set him at work to remove a pile of bricks; the bricks being removed, and the gentleman not yet having another job ready for him, ordered him to remove the bricks back again, to do which he flatly refused; feeling, as a true man should feel under like circumstance, that he was not giving an equivalent for his wages; that his work being of no earthly use to his employer or any one else, it made him a destroyer, and not a producer. His motives were appreciated by the gentleman, who soon found him more important and useful work.

Most men would have discharged the man at once, and would have told him that, if he was willing to pay wages for what was of no use, it was his own business. You may also say that as the gentleman was rich, and could afford to pay for work that was of no use, the man ought to have removed the bricks and taken his wages; for though nothing was produced, yet the world was none the poorer. Let us look at the principle here involved a little closer. Suppose the millionaire, as he could well afford it, had employed fifty men instead of one, at ten dollars a week, for removing bricks from one place to another, the $500 paid for the week's work was lost; no value being added to the bricks, they were worth no more on Saturday night than they were on Monday morning, although $500 had been paid for labor upon them. Is the world no poorer? It may be said the money only changed hands. True; the money changed hands, and perhaps the men spent all or nearly all their wages for food and clothing for themselves and their families. But what has become of the food and clothing and the other materials that were worn out during this week of non-productive labor? Food, clothing, etc., were consumed, and nothing produced. Thus the world was certainly $500 the poorer for the week's unproductive labor. But if these fifty men had been employed to build a house with the bricks, there would have been a house in return for the money; and though the world would have been $500 poorer in food, clothing, etc., yet it would have been the richer by a house.

Though this is an extreme case—for men do not generally pay wages for work that will not be of use to them—yet men very often spend their money for what is not any more productive, but much more injurious, to themselves and society. In the production of malt liquors and all other intoxicating drinks the real sources of wealth, land, and labor are employed unproductively.

MAKING AND SELLING INTOXICATING DRINKS IS UNPRODUCTIVE LABOR.

The labor devoted to brewing, distilling, and the selling of liquors is unproductive labor ; for while the liquors do not benefit those who consume them, yet the necessary and useful products of labor are consumed, or rather destroyed, in the process of making them. What a man consumes to keep up his health, strength, and capacity for labor at some beneficial employment is productive consumption ; but when money is expended for intoxicating liquors, whether by the idle or industrious, it is unproductive consumption in the highest sense of the term. Intoxicating drinks neither preserve health, give strength, prolong life, nor in any way aid the consumer to perform labor ; but they injure health, unfit for productive employment, and will ultimately shorten life. And not only are the millions of dollars wasted that are expended for intoxicating drinks, but the value of the products used, and the labor expended in their manufacture and sale, are lost to society. To these losses must be added the loss of the products of industry consumed by those and their families who are engaged in either the manufacture or the sale of intoxicating drinks ; for all are unproductive laborers.

If those engaged in the liquor-trade are non-producers, and as they are necessarily consumers, it follows that such persons are really no better than paupers—ay, worse than paupers. Their business is not only unproductive, but it retards and prevents productive laborers from pursuing their useful occupations, thus inflicting society with a threefold loss.

THE USE OF LIQUOR, ETC., CAUSES UNPRODUCTIVE CONSUMPTION.

It would be absurd to make laws merely to find employment for lawyers, or to spread disease to give practice to doctors ; and it would be equally absurd for the Government to encourage the liquor-traffic merely to find business for some thousands of liquor-makers and liquor-sellers, to consume the fruits of the industry of others. The man that lives upon the products of another's industry does not create any beneficial demand for those products ; he is merely a destroyer.

The manufacturers and distributers of alcoholic drinks do not create a healthy demand for the products used in the manufacture of the drinks, nor for those consumed by themselves, their families, or employees. The consumption is unproductive, because the strength derived from the food consumed is not used in producing other beneficial things. The blacksmith consumes perhaps as much as the liquor-seller; but the smith returns value in horseshoes and other useful iron-work, while the drink-seller adds nothing to the common weal, but really abstracts by rendering his customers less able and less willing for useful labors.

You may say the money spent remains in the country and forms a part of the general wealth. True; it remains in the country. But what are the general effects of its expenditure? If a member of a family should appropriate to himself the income of the family, and then say to the other members, "What does it matter? it's all in the family," it is very clear that it would not be of much advantage or consolation to those deprived of their right to a share.

All the products of a nation will be consumed sooner or later, and are produced to be consumed; and their value is inert until ready for consumption; and money spent for one article will necessarily preclude the same money being expended for another of equal value. It is less important to gratify artificial wants than those of first necessity; it is, therefore, more important that our people should spend their revenues for what is of utility than for intoxicating drinks, which, as already said, are not only useless, but unfit the consumers for productive labor. We are told by the Brewers' Congress, in a resolution, that they, "from the amount of capital invested in their business, from the labor they employ, and from a large proportion of an article of necessary and general

consumption of what they produce, and more particularly from the steadily-increasing progress made by
their trade, as conclusively shown by the Internal Revenue returns and other statistics gathered from most
trustworthy sources, presented to this Congress, do
represent a very important branch of manufacture,
and are entitled to demand from the National Government, and also from the various State legislatures, proper recognition and protection''! Thus it is plainly
claimed by the brewers, as by others engaged in the
liquor business and those favorable to its existence,
that the liquor-traffic is beneficial, because it employs
a large number of persons and causes a large amount
of capital to be invested.

Though all wealth is the result of labor, yet a great
amount of labor is often expended that does not increase the wealth of the nation, but destroys it. The
latter is eminently the case with labor employed in
manufacturing and selling intoxicating drinks. That
the truth may be clearly seen, we will carefully examine and compare statistics on both sides of the question. To confine ourselves more closely to the subject,
we will take Pennsylvania to represent the United
States.

DOES THE LIQUOR-TRAFFIC CREATE A DEMAND FOR LABOR?

We have no official report of the actual first cost of
intoxicating liquor in Pennsylvania or any other State,
and can only approximate such cost by taking the
general average based upon the sales of licensed

liquor-dealers, which we have estimated at $5,000 each per annum. The indirect and consequential cost of liquors, probably fully equal to the direct cost, is excluded. Assuming this average, then, the 15,745 licensed retail liquor-dealers of Pennsylvania returned by the Revenue Department report of 1872 sold in that year liquors costing the consumers $78,725,000.

This sum is certainly not more than was sold when we consider the great number of unlicensed liquor places in the State, and especially in Philadelphia, where there are (as the writer was informed by a member of the Liquor-Dealers' Protective Association of Philadelphia) more than 4,000 places where liquor is sold without license. This estimate of $78,725,000 is nearly the value of all the woollen goods,* $27,361,897; cotton goods,* $17,565,028 ; boots and shoes,† $16,-864,310 ; furniture and house-fixtures,‡ $9,389,503 ; and all the worsted goods,§ $7,883,038, that were returned by the Census of 1870, as manufactured in the State.

The drink-bill of Pennsylvania is only about twelve million dollars less than the value of all the food and food-preparations,‖ $66,564,919 ; and the manufacture of clothing,¶ $23,363,156, produced in the State in 1870. If the liquor-traffic did not exist, these $78,-725,000 would mostly be expended by our people for the comforts and necessaries of life, which would not

---

* Census Report 1870, Vol. III., p. 489.   † Ibid., p. 563.

‡ Ibid., p. 437.   § Ibid., p. 633.

‖ See Census Report, Vol. III., " Wealth and Industries," for 1870, p. 436.

¶ Ibid., p. 563.

only increase their happiness and virtue, but would in a high degree promote the general prosperity of the State and nation. Let us suppose that our people stop the use of liquor for one year. Now let one-fifth of this drink-bill be deposited in our banks as a reserve fund, and $50,000,000 expended for necessaries and luxuries, as follows:

1. Let one-sixth of the $50,000,000, or $8,333,333⅓, be spent for farm and market-garden products, such as wheat, corn, beef, mutton, poultry, eggs, butter, fruits, and vegetables, which would give our farmers and gardeners twice the profit that they now receive for the fruits and grain manufactured into strong drinks.

2. Let one-tenth, or $5,000,000, be expended in the building of houses. This sum would provide for our working-men 3,571 homes of the value of $1,400 each. To build these houses would give employment to 2,000 persons, and pay them* $1,000,000 for wages, $2,600,000 for building materials, and pay a good interest on $1,400,000 of capital, besides vastly improving the home enjoyments of our people. To prepare the $2,600,000 worth of building materials would give employment to thousands of laborers, as lumber-men, sawyers, etc., etc.

3. Let another one-tenth, or $5,000,000, be expended for cotton goods, which would increase the demand for raw cotton, etc., start the mills now stopped, or cause new ones to be erected, or change the distilleries

* See Census Report, Vol. III., under specific manufactures, on which the above estimates are based.

and breweries into factories to supply the increased demand for cotton goods. It would give employment to 4,000 factory operatives, pay $1,100,000 for wages, more than $3,000,000 for raw materials, and give a good investment of $3,000,000 in factories, etc.

4. Let another one-tenth, or $5,000,000, be spent for woollen goods, which, besides creating a demand for wool, the produce of our own and foreign coun-tries, would give employment to 2,500 factory operatives and others, pay for wages $800,000 and $3,400,000 for raw materials, and give good invest-ment to nearly $2,500,000 capital.

5. Let another one-tenth be expended for worsted goods, which would give employment to 2,500 hands, pay $850,000 for wages and more than $3,000,000 for raw materials, and give investment to $2,000,000 capital.

6. Let another one-tenth be invested in furniture and house-fixtures, which would give employment to 3,500 persons, pay for wages about $1,500,000, pay for materials $2,000,000, and give an investment to $3,000,000 capital.

7. Let one-sixth, or $8,333,333⅓, be invested in clothing, and it would give employment to 6,500 per-sons, pay for wages $1,660,000, and for raw materials pay $4,000,000, and give a good investment to $3,-600,000.

8. Let another sixth, or $8,333,333⅓, be spent for boots and shoes, and it will give employment to 7,650 persons, pay for wages $2,800,000, pay for materials $3,500,000, and give investment to $3,250,000 of capital.

In addition to the above advantages to be derived by our people from expending their money for useful articles and comforts, instead of intoxicating drinks, it will give $12,980,000 for profits and expenses upon the commodities above named after leaving the place of manufacture, and until they reach the consumers.

TABLE IX.

Shows the number of persons that could be employed, the wages paid, value of materials used, the capital invested, etc., by expending but little more than half of what is paid for intoxicating drinks in Pennsylvania.*

Name.	No. Persons employed.	Wages paid.	Cost of Materials.	Capital Invested.	Total Cash Expended.
		Dollars.	Dollars.	Dollars.	Dollars.
Farm products, .	....	....	....	....	8,333,333⅓
Building houses, .	2,000	1,000,000	2,600,000	1,400,000	5,000,000
Cotton goods, .	4,000	1,100,000	3,000,000	3,000,000	5,000,000
Woollen goods, .	2,500	800,000	3,400,000	2,500,000	5,000,000
Worsted goods, .	2,500	850,000	3,000,000	2,000,000	5,000,000
Furniture, etc., .	3,500	1,500,000	2,000,000	3,000,000	5,000,000
Clothing, .	6,500	1,660,000	4,000,000	3,600,000	8,333,333⅓
Boots and Shoes, .	7,650	2,800,000	3,500,000	3,250,000	8,333,333⅓
Total manuf's, .	28,650	9,710,000	21,500,000	18,750,000	41,666,666⅔
Total, . .					50,000,000

By this table it will be seen that by expending for useful and necessary articles of our manufactures only $41,666,666⅔, or little more than one-half of what is spent by the consumer annually for liquors in Pennsylvania, it would give employment to 28,650 hands, pay $9,710,000 in wages, use $21,500,000 worth of raw materials, and find an investment for $18,750,-000 of capital in the manufacture of the articles named, which would certainly increase the happiness, comfort,

* These estimates are based on the Census returns, and are all less than the true amount.

and prosperity of our people, and promote beyond all calculation the general wealth, power, and influence of the State. But it may be said: "The money that is expended for liquors also employs labor, causes the use of materials and the investment of capital." All of which is readily granted. In answer, we have this to say: that the more labor employed and the more capital invested in the liquor business, the worse is the case of the liquor-men; for it is certainly so much labor and capital wasted. But we believe it can be clearly shown that the capital invested in the liquor business employs the smallest amount of labor and confers the least benefit on the nation.

Among the manufactures of Pennsylvania, as given in the Census returns of 1870, are the following:

TABLE X.

Showing the Liquors Manufactured in Pennsylvania in 1870.

Kind of Liquors.	Hands employed.	Wages paid.	Cost of materials.	Capital invested.	Value of the Liquors.
		Dollars.	Dollars.	Dollars.	Dollars.
*Distilled liquors, .	512	215,837	1,950,077	2,504,857	4,618,228
*Fermented liquors,	1,583	773,267	3,553,986	6,966,236	7,056,400
*Vinous liquors, .	15	4,250	7,960	100,160	17,900
Totals, . .	2,110	993,354	5,512,023	9,571,253	11,692,528

From the Census returns of 1870, the whole of the liquors that were manufactured that year employed 2,110 hands; paid for wages, $993,354; used $5,512,023 worth of materials; there was invested $9,571,253; the value of liquors was $11,692,528 at the place of manufacture.

* Census Report, Vol. III., p. 451.

Let us now compare the totals of Table IX. and the totals of Table X., and see how the question stands :

Kind.	Persons employed.	Wages paid.	Cost of materials.	Capital Invested.
		Dollars.	Dollars.	Dollars.
Totals of Table IX. of useful articles, ...	28,650	9,710,000	21,500,000	18,750,000
Totals of Table X. of liquors, ...	2,110	993,354	5,512,023	9,571,253
The difference, . . .	26,540	8,716,646	15,987,977	9,178,747

By the difference of totals we find that the money, if spent for useful articles, would employ 26,540 more hands ; pay $8,716,646 more for wages ; pay $15,987,-977 more for materials ; and invest $9,178,747 more capital to produce $41,666,666⅔ worth of useful and necessary articles, than it would to produce $11,692,-528 worth of liquors at the places of manufacture.

It may be objected that there is a difference in the value of the products at the place of manufacture, and that a difference in amount of capital, labor, etc., should be expected, and that to form a just comparison the value of products should be nearly the same. What we have mainly to do with is the cost to the consumers. Let us, then, endeavor to ascertain the cost of the two classes of commodities to the consumers, and then their relative effects on labor, capital, etc.

In 1870, by the Internal Revenue Report, there were manufactured in Pennsylvania spirits and malt liquors as follows :

Distilled liquors of all kinds, 5,361,920 gals., worth at retail $6 a gal.,			$32,171,520
Fermented liquors, . . 783,034 barrels, " " $24 a bbl.,			18,912,816
Total cost at retail to consumers, ...			$51,084,336

This $51,084,336 is certainly less than the cost of liquors in Pennsylvania; for our people as a rule are not more sober nor drink less than the average of the people in the other States of the Union. The people of Pennsylvania pay about 6 per cent. of the revenue collected on the spirits manufactured in the United States, and about 12 per cent. of the revenue on malt liquors. Unless our people drink less than their share of the liquors produced and consumed in the country, more liquors are consumed in Pennsylvania than are annually manufactured in the State.

If to the money expended for building houses and for the purchase of cotton goods, woollen goods, worsted goods, furniture, clothing, boots and shoes, as given in Table IX., we add 25 per cent. on the money so expended for profits, expenses, etc., from the place of manufacture until they reach the consumers, then the cost to consumers of these articles of first necessity will be $52,083,332, while intoxicating drinks, as we have seen, cost $51,084,336, or only $998-996 less than the cost of the 3,571 houses and all the manufactures specified in the table, which, as before said, would give employment to 26,540 more hands; pay $8,716,646 more for wages; pay $15,987,977 more for materials; and invest $9,178,747 more capital to produce them than to produce the intoxicating drinks. Now, if it is good political economy for a people to encourage that which will cause the greatest amount of labor and employ the most capital, without any regard to the results or the products of the labor and capital employed, it is certainly good political economy for our people to discourage the expenditure of money for liquor.

But when we consider the utility of houses, cotton goods, woollen goods, etc., etc., with the injurious effects produced by intoxicating drinks, every argument in favor of the use and the traffic in strong drinks becomes drivelling, if not devilish.

Is it not clear, then, that the liquor-traffic causes bad trade, employs the least amount of labor for the money spent, to say nothing of its inutility?

We will show, as clear as facts and figures can show, in the next chapter, that the drinking system is not only injurious to trade and commerce, but to all useful labor.

CHAPTER VII.

THE drink-bill of the United States, as we have shown, for the four years of 1869–70–71 and '72, is $2,729,186,709, or an average of nearly $700,-000,000 a year, which is a less sum than the average for the years since 1872. Experience and observation have fully demonstrated that when men cease to use intoxicating beverages they invariably devote a very large portion of the money formerly spent for drinks to the purchase of useful articles for themselves and families. It is therefore very reasonable to conclude that if the liquor-traffic was swept from our country, not less than three-fourths of the money now spent for liquors would be devoted to the purchase of useful articles.

Let us suppose that our people, instead of spending for drinks $700,000,000 annually, applied that sum to the purchase of the following, being one-half of the articles named produced in the United States, as given in the Census of 1870, viz. :

I. Div.* "For food and food-preparations," . . $300,182,785
II. Div.† "Cotton goods," 84,228,676
III. Div.‡ "Woollen goods," 75,649,098
IV. Div.§ "Boots and shoes," 90,822,045
V. Div.‖ "Furniture and house-fixtures," . . 37,769,859
                                                _____

    Total, ... $588,652,463
Which will leave for expenses, profits of retailers,
    etc., ... 111,347,537
The total being less than the direct annual cost of   _____
    liquors, ... $700,000,000

Let us see what necessaries or useful articles these seven hundred million dollars would purchase in the place of the 272,530,105 gallons of poisonous liquids now annually consumed.

I. Div.—For the first item of the above general division, viz., $300,182,785, devoted to the purchase of "food and food-preparations," let us specify what could be purchased by a proper expenditure of this portion of our drink-waste.

1. We could purchase 16,039,572 ¶ barrels of wheat-flour, 222,125 barrels of rye-flour, 15,596,981 bushels of corn-meal, 202,230 cwt. of buckwheat flour, 280,375 bushels of hominy, and 14,703,732 cwt. of cattle-feed.

2.** These grist-mill products would require to manufacture them, as we learn from our last Census Report, 11,286 mills, of the average capacity of grist-mills in the United States, give employment to 29,224 persons, and pay $7,288,766 for wages.

* Vol. III. Census Report, 1870, p. 435.
† Ibid., p. 430.   ‡ Ibid., p. 489.   § Ibid., p. 416.   ‖ Ibid., p. 437.
¶ The quantities of the articles above named are proportional quantities of the same as given in Vol. III. (Wealth and Industries), Ninth Census of the United States, 1870. The figures are actual results as furnished by the official tables, except in a few instances, when explanations and reasons are given. In many, if not in most, figures of quantities, those given are less than the official data would warrant.
** Vol. III. Census Report, 1870, pp. 598, 599.

3.* The following farm-products or raw materials would be used in the above mill manufactures, viz. : 183,274,484 bushels of wheat, 20,498,226 bushels Indian corn, 9,802,488 bushels of oats, 1,110,625 bushels of rye, 444,461 bushels of buckwheat. The value of the grain used would be $181,157,263, and the value of the flour and other grist-mill products $222,492,571.

4.† And to prepare these products for family use would require 1,775 bakeries, of the average capacity, give employment to 7,063 persons, and pay them for wages $2,676,592.

It is evident that such an increase of bread and breadstuffs would not only benefit the farmers, grist-millers, flour-merchants, bakers, etc., but every other useful trade in the country, and also increase the general health. The history of all nations has fully established the fact that, when bread and breadstuffs are plentiful and the people well supplied, the general health is good and mortality lessened. While this increased consumption of bread and breadstuffs would benefit our millers, bakers, merchants, etc., the augmented demand for the products of our farmers would enable them to improve their farms and increase their crops and the wealth of the nation.

The benefits of total abstinence from alcoholics, and the non-existence of the traffic in them, would further encourage our agriculturists : 1. by increasing the demand for cattle for ‡ slaughter to the value of about $5,000,000 annually, which would employ 941 persons as butchers, pay $273,173 for wages, and cause an increase of the sales of meats to value of not less than $5,519,064; and thus, while it supplied an abundance of wholesome food, would increase the general business of the country.

5. It would also give to our * farmers and graziers $775,000 for animal food-preparations, $1,265,776 for beef and pork cured and packed, $762,000 for beef packed, $28,000,000 for pork packed, and $1,500,000 for market-garden and orchard products canned and preserved.

6.† Besides the benefits that would flow from the increased demand for the above products of our agriculturists, the preparation of these products for market would employ 249 persons in preparing cured and packed meats, and pay $86,590 for wages; to pack beef, 217 persons, and pay $55,797 for wages ; to pack pork, 2,775 persons, and pay for wages $861,163 ; and to can and preserve fruits and vegetables, would employ during the season 2,934 persons, and pay $385,821 for wages.

7. ‡ There would be dairy products, consisting of 52,233,202 gallons of milk, worth $7,040,142, made into 54,767,614 lbs. of cheese, worth $8,355,284, and $30,548 worth of other products. Total value of products of cheese-factories was $8,385,832, which would employ 2,303 persons, and pay for wages $353,283.

8.‡ Nor do the advantages of abstaining from intoxicating drinks end here ; for in the preparation of coffee, spices, etc., there would be employed 610 persons ; pay them for wages $335,491, pay for materials $4,085,918, producing coffee, spices, etc., for market worth $5,633,211.

9.* For fish cured and packed there would be employed 427 persons ; pay them for wages $90,552, pay for materials $420,602, and the products, when ready for market, would be worth $796,295.

10. Fish and oysters canned employ 793 persons ; pay for wages $133,204, use materials worth $470,151, the value of products $735,650.

11.* Food-preparations of vegetables employ 279 persons ; pay for wages $88,531, use $373,449 worth of materials, the products worth $593,101.

12.* The preparation of ground mustard, preserves and sauces, vermicelli, and macaroni, and chocolate employs 334 persons ; pays $126,156, uses materials worth $876,045, and the value of the products is $1,301,837.

There cannot be the least doubt but that the demand for the above-named food and food-preparations, or products equivalent to them, would be increased in about the proportion estimated, if our people would abstain entirely from intoxicating drinks for one year.

This increase of business, with its general and special beneficial results, would not end with those named ; for the increase of one branch of *productive*

* See report of articles named, Vol. III.

industry will increase others, whether directly connected with it or not.

There is a reciprocal connection between all *productive industries*, except when articles of like use are substituted one for another, as cotton goods for woollens, etc., etc. But between the products of labor given above and intoxicating drinks there is for ever an antagonism. · The use of these drinks is at war with all the productive industries and the labor interests of the nation.

TABLE XI.

Shows by expending $300,182,785 for food and food-preparations how many industries will be promoted, the number of persons employed, wages paid, value of materials used, and value of each product.

INDUSTRIES.	Estab-lishm's.	Pers'ns em-ployed.	Wages paid.	Value of ma-terials used.	Value of products.
			Dollars.	Dollars.	Dollars.
Flour and grist mill products,*. . .	11,286	29,224	7,283,766	183,696,061	222,492,571
Bread, cracker, and bakery products, .	1,775	7,063	2,676,592	11,105,928	18,453,852
Butchering, . . .	254	940	273,173	5,519,964	6,843,080
Coffee, spices, roast'd, etc.	78	610	335,401	4,085,918	5,633,211
Fish, cured and packed.	37	427	90,552	420,602	796,295
Fish and oysters, can'd.	8	793	133,204	470,151	735,650
Food preparat'n, anim'l.	42	291	138,218	774,240	1,164,395
" vegetable,	16	279	88,531	373,449	593,101
Fruits and vegetables, canned and preserv'd,	48	2,934	385,821	1,547,423	2,712,838
Meat, cur'd and pack'd,	8	249	86,590	1,265,776	1,880,401
Packed meat, beef,	18	217	55,797	762,340	975,153
Pork, packed, . .	103	2,775	861,163	23,288,932	28,214,665
Ground mustard, . .	7	47	21,543	99,417	153,504
Preserves, sauces, etc.,	15	167	56,713	422,385	621,418
Cheese, . . .	656	2,303	353,283	7,044,642	8,385,832
Vermicelli and macar'i,	3	18	8,550	21,669	53,736
Chocolate, . . .	4	99	39,350	332,574	473,129
Totals,† . . .	14,358	48,436	12,893,337	241,231,471	300,182,781

* Census Report, Vol. III., pp. 391, 395, 396, 397.   † Ibid, p. 435, each one-half.

By the foregoing table we see that our people, by abstaining from alcoholic liquors, and by spending less than one-half of their annual cost, $300,182,781, for food and its preparations, would need 14,358 additional food-preparing establishments, employ 48,436 persons to carry on these manufactories, pay them for wages $12,893,337, and cause a demand for $241,- 231,471 worth of raw materials. There is no exaggeration in these figures, nor are they made for the purpose, but are proportional estimates from the Census Report of 1870, with all the fractions thrown out of the computation. Considering the thousands of our people who are under-fed or in an actual starving condition, we cannot doubt but that this sum of money would be spent for food, if not expended for drink. Our farmers would receive for grain alone used for these food-preparations $181,157,263, which is $131,056,975 more than the value of all the materials used in the manufacture of liquors in 1870; for the Census returns give $49,100,288 * as the value of the materials made into liquors.

THE BENEFIT OF MONEY SPENT FOR MANUFACTURES.

Having seen some of the benefits that would result from spending a portion of our national drink-bill, viz., $300,182,785, for "food and food-preparations," we will endeavor to point out the advantages to be derived from spending the other portions for some other of our manufactures.

II. Div.—Let us see what would be received by

spending $84,228,676 for "cotton goods," and not for intoxicating drinks.

1:* We should receive 239,102,256 yards shirtings, sheetings, and twilled goods, 17,266,731 yards of lawns and fine muslins, 244,625,026 yards of printed cloths, 15,150,543 lbs. yarn not woven, 5,780,121 dozen spools thread, 5,559,063 lbs. of bats, wicking, and wadding, 36,509,022 yards of warps, 246,946 table-cloths, quilts, counterpanes, 1,383,530 seamless bags, 2,528,727 lbs. of cordage, lines, and twines, 4,195,025 yards of flannel, 453,034 lbs. thread, 19,-637,622 yards ginghams and checks, 6,970,447 yards of cassimeres, cottonades, jeans, etc., and several million pounds of other cotton fabrics.

2.* In the production of these cotton goods there would be consumed or used 199,154,128 lbs. of raw cotton, 3,111,094 lbs. cotton yarn, 68,050 lbs. cotton warp, and 2,617,130 lbs. cotton waste ; the cost of mill-spindles $5,455,336, and all materials $55,868,468.

3. To produce these cotton goods would require 478 factories of the average capacity of the United States establishments, employ 67,684 operatives, and pay them $19,522,066.

4. The benefits do not consist alone of those already named ; for it must be remembered that the nearly two hundred million pounds of raw cotton are the products of our agriculturists of the Southern States ; besides, the other materials used in the manufacture of these cotton goods are the products of our

* Census Report, Vol. III., pp. 596, 597.

people's industry, not included in the above numeration.

III. Div.*—Then, again, let us suppose that we invest another portion of the drink-bill, say $77,702,679, in "woollen goods,"what would be the result of this investment?

1. To minister to health, comfort, and enjoyment, there would be received 1,000,219 pairs of blankets, 130,604 yards beavers, 31,670,306 yards cloth, cassimeres, and doeskins, 970,932 yards felted cloth, 966,191 yards of negro cloth, 37,500 yards cottonades, 113,372 coverlids, 29,482,643 yards flannels, 37,500 yards of frockings, 12,244,992 yards jeans, 2,753,451 yards kerseys, 7,065,137 yards linseys, 1,331,883 yards repellants, 7,036,279 yards satinets, 140,000 yards Balmoral skirts, 1,426,729 yards tweeds and twills, 61,000 lbs. of warp, 7,078,118 lbs. of yarn, 111,500 lbs. of hosiery yarn, 784,500 pounds of shoddy yarn, 10,730 dozens of hosiery, 4,341,534 lbs. of rolls, and for men and horses 29,276 blankets.

Would not these products of our woollen factories be a good exchange for the millions of gallons of the poisonous decoctions made in the breweries and distilleries?

2. The benefits do not end with the consumers of these goods, but their results will extend to the producers. For to manufacture these goods would require 1,445 factories, the average size of the factories in the United States; to drive the necessary machinery, it would take at least 525 steam-engines and

* Ibid., pp. 630, 632.

1,046 water-wheels of the aggregate power of 47,616 horse-power; of machinery, it would need 4,183 sets of cards, capable of carding daily 428,696 lbs. of wool, 7,019 broad looms, 10,072 narrow looms, and 922,748 spindles ; and besides giving work to men to make the machinery, it would give employment to 40,026 persons, and pay them for wages $13,438,787.

3. Nor would the results end here ; for in the manufacture of these goods there would be used 9,686,031 lbs. of shoddy, 656,280 lbs. of warp cotton, 77,383,547 lbs. of domestic wool, 8,655,912 lbs. of foreign wool, 1,631,974 lbs. cotton yarn, 1,286,759 lbs. woollen yarn, $2,916,673 worth of chemicals and dyestuffs, and $2,835,125 worth of other materials. The total value of the materials to produce these goods is $48,216,300.

IV. Div.*—In addition to what has already been enumerated, total abstinence would allow our people to expend $90,822,045 for boots and shoes.

1. For which we should receive 7,159,264 pairs of boots, worth $25,115,735 ; and 33,154,357 pairs of shoes, valued at $46,923,103, and other articles worth $1,-305,589.

2. There would be used in making them 2,466,201 sides of sole leather, 3,807,356 sides of uppers, and 6,892,721 lbs. of other leather. To make them would require 67,944 hands, and pay them for wages $25,-986,356, and use materials worth $46,791,264.

3.† To produce the shoe-findings would require 135 establishments, employ 1,386 hands, paying for wages $396,478.

* Ibid , p. 591.   † Ibid., pp. 415, 416.

4.* To tan, curry, and prepare the leather would need 3,784 establishments, and employ 17,621 persons, and pay $7,252,887 for wages.

V. Div.—There could also be expended for all kinds of house-fixtures and furniture (except stoves and hollow-ware) $37,769,859.

1.† To produce which would need 3,156 establishments, employ 28,545 persons, and pay for wages $11,652,478, and for materials $14,258,272.

2. In addition to the hands employed in making these house-fixtures and furniture, employment would be given to thousands of other persons, as lumbermen, sawyers, etc. This expenditure, no one will deny, would materially add to the comforts and the enjoyments of our people. It would not only make the homes of our laboring classes more pleasant and desirable, but would cultivate a greater love for home and home enjoyments. This would tend to form better habits among our rising generation, to develop moral and religious sentiments, and to form a superior class of citizens for the future of our country, which would redound to the happiness and prosperity of our people, and to the honor and glory of our nation.

By the totals of Table XII. it is seen that by expending $288,469,678 for cottons, woollens, boots and shoes, furniture, etc., instead of intoxicating drinks, there would be required 20,712 additional manufacturing establishments, employ 223,206 hands to run them, pay for wages $78,249,052, and for materials $165,134,304.

TABLE XII.

Exhibits the number of additional establishments that would be required, the hands employed, the wages paid, the value of material used, and the value of the products, if $288,496,678 were expended for the articles named.

Manufactures.	Establishments.	Number employed.	Wages paid.	Value of the materials.	Value of products.
			Dollars.	Dollars.	Dollars.
Cotton goods, . . .	478	67,684	19,522,066	55,868,468	88,744,869
Woollen goods, . .	1,445	40,026	13,438,787	48,216,300	77,702,679
Boots and shoes, .	11,714	67,944	25,986,356		
Shoe-findings, .	135	1,386	396,478	46,791,264	90,822,045
Tanning, Currying Leather, . .	3,784	17,621	7,252,887		
Fur'ture and h'use-fixtur's,	3,156	28,545	11,652,478	14,258,272	37,769,859
Total of manufactures, .	20,712	223,206	78,249,052	165,134,304	295,039,452

TABLE XIII.

Exhibits the totals of Table XI. and Table XII., and the aggregate of the totals.

	Establishments.	Hands employed.	Amount of wages paid.	Value of materials used.	Value of the products.
			Dollars.	Dollars.	Dollars.
Totl's of Table XI., food and food-prep'r'tions,	14,358	48,436	12,893,337	241,231,471	300,182,781
Totals of Table XII., of manufactures, . .	20,712	223,206	78,249,052	165,134,304	295,039,452
Aggregate, . .	35,070	271,642	91,142,389	406,365,775	595,222,233

Thus our people, by expending $700,000,000, our average annual drink-bill, for the productions given on Tables XI. and XII., would keep running 35,070 establishments, employ 271,642 persons, pay for wages $91,142,389, and cause a demand for $406,365,775 worth of raw materials, and leave $104,777,767 for profits and expenses upon the commodities after leaving the place of production and until they reach the consumers.

Will any person who is able to reflect or to distinguish between a benefit and an injury maintain that the nation would not be benefited beyond all calculation by the prohibition of a traffic which prevents so much productive labor and wastes so much of our nation's wealth for that which is not only useless but positively injurious to the consumers individually and to our people collectively ?

It is clear, in view of what has been already said and the facts and figures presented, that the government which acts so absurdly and irrationally as to license the sale of intoxicating drinks violates sound principles of political economy. The traffic in, and use of, these drinks not only prevents productive labor and wastes the capital expended for them, but the grain, fruits, and other materials used in their manufacture is lost ; and all the people of the nation, whether they drink the liquors or not, have to make up or suffer the loss, and they who drink the liquors not only lose all the capital they expend for them, but have to bear their share also of the general loss.

No one derives benefit from the liquor ; the consumer, from its effects, suffers a loss of physical and mental power, and he would have been the gainer in health and power if he had cast his money into the fire ere he spent it for the drink.

There is not a shadow of doubt but that more than *seven hundred million* dollars would be annually spent for food, clothing, and other articles of use and comfort, additional to what is now expended, but for the use and the traffic in strong drinks. This additional sum expended, for our manufactures and products of agriculture annually, would give a great impetus to every department of industry, manufactures, agriculture, trade, and commerce. From the facts and figures presented it must be plain that the remedy for our present bad trade and lack of labor lies entirely within the power of our people. If we continue to spend these more than seven hundred million dollars annually for poisonous drinks, and expect to have prosperous trade, we shall find out our mistake when it is perhaps too late to apply the remedy. As the use of the drink prevents productive industry, it is logically clear that we may increase productive labor and trade by stopping the drink and removing the temptations to drunkenness.

CHAPTER VIII.

THE USE OF, AND THE TRAFFIC IN, STRONG DRINKS INJURES LABOR.

THE demand for any product creates a demand for labor to produce it; and when there is no demand for an article, there is certainly no need for its manufacture. It is equally true that wages are regulated by the demand for labor. When the number of laborers exceeds the demand for their labor, wages will be low; and when the laborers are less than the demand, wages will be high. In other words, when two men are seeking the same job of work wages will be low; but when two jobs are seeking one man, wages will be high. It is equally clear that when the demand for labor just equals the supply, the working-classes can obtain "a fair day's wage for a fair day's work" —that "Labor will not be oppressed by Capital." But you ask, "How can this employment and fair wages be obtained?" Nothing is more simple. All that needs be done is to create a demand for the useful products that will give the most labor and extend their beneficial influences to promote other productive industries. Let the millions now wasted, or worse than thrown away, for liquors, be spent for food, clothing, furniture, and other necessaries, and not

only would there be work for our unemployed and
those engaged now in the liquor business, but labor
would be in demand in all departments of productive industry. Not only does the money expended
for liquors give the least amount of labor in proportion to the capital swallowed up in their purchase,
but experience proves that, while the user of them is
impoverished, those that it employs are also most
generally debauched and ruined.

The census returns in 1870 show that the value of
the malt and spirituous liquors manufactured in
Pennsylvania were valued at the place of production
at $11,692,528; that their manufacture gave employ-
ment to 2,110 hands, and paid $993,354 for wages;
which, on an average, is less than eight and one-half
cents' worth of labor to make one dollar's worth of
liquor. A gauger in the service of the United States
to whom the writer applied for information as to the
cost of manufacturing liquors writes: "At present
I have but one distillery in my charge. It employs
three men at an average pay of $2 25 per day each;
uses 48 bushels of grain, producing, on an average,
188 proof gallons of spirits per day, which, in its
crude state to-day (Sept. 6, 1873), is selling at $1 05
per gallon. To the cost of this quantity of grain (48
bushels) daily add the cost of two and one-half
pounds of hops, three-fourths of a ton of coal, and
three empty barrels." As this distillery is in Pennsylvania and in a neighboring county, we will take
this as the average cost of manufacturing spirits in
Pennsylvania.

The cost of manufacturing 188 gallons of crude rye whiskey will be as follows:

Rye, 48 bushels, at $1, ... $48 00
Hops, 2½ lbs., at 40 cents, ... 1 00
Coal, ¾ ton, at $6, ... 4 50

Total cost of material, ... $53 50
Three men's wages, at $2 25 per day each, . . . 6 75
Three empty barrels, at $2 each (iron-bound), . . 6 00

Total cost of materials, barrels, and labor, $66 25

188 gallons of whiskey (crude), at $1 05, . . . $197 40

Thus the cost of labor to manufacture 188 proof gallons of crude whiskey, worth at the distillery $197 40, is $6 75, or less than 3.42 per cent. of the value of the product for labor.

At this rate the 5,361,920 gallons of whiskey made in Pennsylvania in 1870, at $1 05 per gallon in its crude state at the distillery, would be $5,630,016, and at 3.5 per cent. for wages is $197,050. This value of the liquor is $1,011,788 more than the value given in the census returns (Table X.), and $18,787 less than the amount paid for wages. To give the liquor-traffic the benefit of all doubts, we will take the figures in the census returns; for it cannot be said we made them suit to our cause, as the figures in census returns were given by the liquor-men themselves to the census-takers; if they are not correct, they alone are to blame.

Table XIV. exhibits some of the leading industries of Pennsylvania, showing the number of hands

employed, the amount of wages paid, the value of
materials used in their manufacture, the capital
invested, and the value of the products at the place
of their manufacture; also the per cent. paid for
labor on the value of the products.

TABLE XIV.

Kind of Product.	Hands Employed.	Wages Paid.	Cost of Materials.	Capital Invested.	Value of Products.	Per ct. for labor on value of product.
		Dollars.	Dollars.	Dollars.	Dollars.	
All industries............	319,487	127.976,594	421,197.673	406,821,845	711,894.344	*17.97
Boots and shoes	15,799	4,818,902	6 932,726	6,375,943	16,864.310	28.5
Clothing.................	19.136	5,040,272	12,822.465	10,378,443	23,363.156	21.57
Furniture, etc............	6,350	2,775,026	3,355.908	5,686.553	9.389,503	29.55
Hardware................	951	403 597	722,863	951,850	1.537,687	26.24
Cotton goods.............	12,762	3.510.534	10 749,472	12,575,821	17,565,028	19.98
Woollen goods...........	12,578	4.340,06·	17.325,849	14,066,785	27 361,897	15.86
Worsted goods...........	3,868	1.363,334	4.932,940	3,350,078	7,883,038	17.30
All kinds of Liquors ...	2,110	993,354	5,512,023	9,579,253	11,692,528	8.5

In the column under per cent. paid for labor it
will be seen that for labor paid in all industries in
the State combined there is paid 17.97 per cent. of the
value of all manufactured articles; for boots and
shoes, 28.5 per cent. ; for clothing, 21.57 per cent. ;
for furniture, house-fixtures, 29.55 per cent. ; for
hardware, 26.24 per cent. ; cotton goods, 19.98 per
cent. ; for woollen goods, 15.86 per cent. ; worsted
goods, 17.30 per cent. ; while for the manufacture of
liquors only 8.5 per cent. is paid for labor. The percentage of the value of materials to the average value
of all the products of the industries of the State is 59
per cent. ; whilst the per cent. of the value of the

* Per cent. for labor on all the manufactures of the State.

materials manufactured into liquors to their value where made is only 47 per cent.

Thus the sum paid for labor to make $100 worth of all the manufactures in the State averages $17 97 ; for boots and shoes, $28 50 ; for clothing, $21 57; for furniture, etc., $29 55 ; for hardware, $26 24 ; for cotton goods, $19 98 ; for woollen goods, $15 86; for worsted goods, $17 30; while for making $100 worth of liquors only $8 50 are paid for labor.

Nor does this show the true injury done to labor by the sale and use of intoxicating drinks ; for we must bear in mind that the cost of liquors to the consumer is proportionately much more after they leave the place of manufacture than the other pro- ducts of our industries. It has been shown at page 88 that the liquors made in Pennsylvania in 1870 would cost the consumers $51,084,336 ; the percent- age of the cost of the liquors to the consumers for labor being only 1.94 per cent.

It will be a fair estimate if we allow an average of 25 per cent. for the increase in value, or the price upon all the products of industry, after leaving the manufactory, to the time when received by the con- sumers.

Table No. XV. shows the value of the articles named at the manufactory ; the increase of value at 25 per cent. ; the cost to the consumers ; the wages paid on them for labor; and the per cent. paid for labor on the cost of the articles to the consumers, or the sum that is paid for labor out of every $100 spent by the consumers.

TABLE XV.

The kind of Products.	The value of the product of the manu-factory, 1870.	The increase of value at 25 per ct. from Producer to Consumer.	The cost of Articles to the Consumers.	The wages paid for Manufactu-ring the Product.	The sum paid for labor out of every $100 worth
	Dollars.	Dollars.	Dollars.	Dollars.	Dollars.
All Industries......	711,894 344	177,973.586	889,867,980	127,976,594	14 38
Boots and Shoes...	16,864,310	4,216,077	21 080,387	4,818,902	22 85
Clothing........	23,363,156	5,840,789	29,203,945	5,040,272	17 25
Furniture and House fixtures } ..	9,389,503	2,347,375	12,184,254	2,775,036	22 76
Hardware.........	1 537,687	359,421	1,922,109	403,597	20 99
Cotton goods.....	17,565 028	4,391,257	21,956,285	3,500,534	15 94
Woollen goods. ...	27 361,897	6 840 474	34,202,371	4,340,066	12 93
Worsted goods. ...	7,883,034	1,970,759	9,853,797	1,363,334	13 83
Liquors...........	11,692,528	....	51,084,836	993,354	1 94

By examining this table it will be seen, by buy-
ing $100 of the aggregate manufacture of the
State $14 38 of it goes to labor for producing it.
For every $100 spent for boots and shoes, $22 85 goes
for labor ; for clothing, $17 25 ; for furniture, house-
fixtures, etc., $22 76 ; for hardware, $20 99 ; for cotton goods, $15 94 ; for woollen goods, $12 98 ; for
worsted goods, $13 83 ; while $100 spent for liquors
will only give to labor $1 94. If we average a day's
work to be worth $2, then one hundred dollars ($100)
spent for boots and shoes will give one work for $11\frac{85}{200}$
days ; for clothing, $8\frac{25}{40}$ days ; for furniture, etc.,
$11\frac{76}{200}$ days ; for cotton goods, $7\frac{97}{100}$ days ; for woollen
goods, $6\frac{49}{100}$ days ; for hardware, $10\frac{99}{200}$ days ; and for
worsted goods, $6\frac{83}{200}$ days ; while one hundred dollars
expended for liquors will only give one man less
than a day's work.

Is it not clearly evident that the use of strong

drinks injures labor and consequently our laboring classes?

The liquor business employing but little labor, and sharing with labor a very small portion of its profits, is for ever at war with all the interests of labor and the working-classes of all countries, as well as with all efforts for the intellectual, moral, and religious advancement of our race.

Considering the facts and figures already presented, it is certainly the interest of our working-classes, if they desire to improve their own condition and that of their fellow-laborers, to use all their power and influence, social and political, to banish the drink-traffic from our land. Every dollar spent for liquor robs labor of nearly a half-day's work ; or, taking the whole liquor waste of $700,000,000, an aggregate annually of not less than three hundred million days' work.

It has been plainly demonstrated that the man who spends a dollar for liquor receives nothing of value ; labor receives less than two cents from the dollar so spent. If a dollar is spent for a pair of shoes for a child, labor would have received nearly 23 cents as its share, instead of less than two cents, as when spent for liquor ; the child would have a pair of shoes, and the man minus a headache. Hence, in examining the question of capital and labor, the drink question is a very important element, which must be duly considered. The remedy for bad trade is certainly in our own hands. As long as men spend their money for liquors which give but little profit to labor, whilst at the same time they take the place of those commodities that give more employment to the laboring classes, we shall have a continuance of hard times, a scarcity of work, and consequently low wages, and the laborer will continue the "slave of capital."

CHAPTER IX.

In addition to the loss of money expended for liquors there is a series of losses that are the inevitable results of such expenditure.

The first of these is the loss of the labor of those engaged in the manufacture and sale of the drink, which we shall endeavor to approximate. By turning to Table VI. it will be seen that in 1872 there were 7,276 licensed wholesale liquor establishments. If there are three persons employed in each, there are engaged in those places 21,828 persons. There were also in that year 161,144 persons licensed to sell liquor by retail. If two persons are employed in each of these liquor-shops, then 322,288 persons are so employed.

Experience shows that there are nearly as many unlicensed liquor-shops as licensed; but suppose there are only one-half so many, there will be employed in the unlicensed liquor business 161,144 persons. We thus have engaged in the United States about 505,260 persons selling liquors.

There were 3,132 distilleries. If five men were employed in each, then they employed 15,660 men.

The Brewers' Congress, June 3, 1874, said that there were employed 3,566 men in malt-houses and

11,138 in breweries. To which we may add 10,000 more persons who were employed about distilleries and breweries as teamsters, blacksmiths, coopers, etc.

The number of persons directly employed in the liquor business may be estimated as follows:

NUMBER OF PERSONS ENGAGED IN MAKING LIQUORS.

Persons engaged in breweries, ...	11,138
Persons engaged in malt-houses, ...	3,566
Persons engaged in distilleries, ...	15,660
Persons variously employed about breweries and distilleries, ...	10,000
Total, ...	40,364

NUMBER OF PERSONS ENGAGED IN SELLING LIQUORS.

In wholesaling, ...	21,828 .
In licensed retailing, ...	322,238
In unlicensed retailing, ...	161,144
Total engaged in selling, ...	505,260

The total number of persons employed in making and vending intoxicating drinks was 545,624; therefore, as we have seen that the wealth of the nation is the result of productive labor, what real benefit to society is the labor of these 545,624 men, even if we leave entirely out of the consideration all the moral aspects and results of their business? None whatever! Their labor is a total and direct loss. Their labor is unproductive, and whatever they consume is unproductive consumption; and, as said in a previous chapter, they are little better than dependants living upon the industry of producing classes. In truth, they are worse than paupers; for their labor is not only unproductive in itself, but prevents productive industry by unfitting the productive laborer who consumes the drink for useful employment.

Each person now engaged in the liquor business, if employed in some branch of useful industry, would be contributing his share to the aggregate wealth of the nation, which would, at the present time, be worth at least $2 a day, or $500 a year, allowing the balance for loss of time by sickness and other causes. This increase of producing labor would add to the wealth of the nation $1,091,248 per day, or $272,812,000 per annum. This more than two hundred and seventy-two million dollars is only a small part of the direct loss annually sustained by the nation in the shape of labor taken by the liquor trades from productive industry.

It is estimated that there are 600,000 drunkards in the United States, which is certainly no exaggeration; for if each of the 161,144 licensed liquor-shops have four customers who are drunkards, the number will be 644,576. That this is a low estimate must be evident when we consider the vast number of un-licensed liquor-shops in the nation. If, on an average, these 600,000 drunkards lose but one-half of their time by drinking, it will equal the loss of the labor of more than 300,000 men annually, which, at $500 a year for each, will be a loss of $150,000,000.

Taking the number of our population at 38,558,371, as at last census, nearly one-half of whom are females, and 8,425,941 males of twenty-one years and upwards, and supposing that only one-sixth of these adult males use intoxicating drinks to any great extent, it will give us for the United States 1,404,323 male tipplers, to say nothing of the female population, though every one knows that they are not all total abstainers. It is estimated by good authorities that there are not less than two hundred thousand female drunkards in the United States. If one day's labor is lost a week by each of these 1,404,323 male tipplers and occasional drunkards, the loss will be $2,808,646 a week, or $146,049,592 a year.

RECAPITULATION OF LOSSES OF TIME AND INDUSTRY.

The loss of time and industry of 545,624 men engaged in making and selling liquor, . .	$272,812,000
The loss of time and industry of 600,000 drunkards,	150,000,000
The loss of time and industry of 1,404,323 male tipplers, ...	146,049,592
Total loss of time and industry, ...	$568,861,592

Investigation will show that this large aggregate is far below the true loss.

But the above are not the only losses which the drink-trade imposes, as will be seen by the following exhibits.

DESTRUCTION OF GRAIN, ETC.

Brewers and distillers destroy grain to produce a product that is unfit to nourish the animal system, while the miller prepares a true food. By the manufacture of liquors not less than 40,000,000 bushels of nutritious grain are annually destroyed. A bushel of rye or corn weighs 56 lbs., and a bushel of barley 47 lbs. The average weight of the grain used for liquors will be about 53 lbs. to the bushel, yielding not less than 40 lbs. of flour, which will make about 60 lbs. of bread, or fifteen 4-lb. loaves per bushel. The 40,000,000 bushels will give a grand total of food annually destroyed equal to 600,000,-000 4-pound loaves of bread, or annually more than 79 loaves for each family in the United States. This calculation does not include the destruction of grain and fruit involved in the manufacture of the liquors imported from foreign countries, nor the domestic liquors produced in the country that are not reported to the Government.

These loaves, if used as paving-stones, would pave a street ten yards wide and more than a thousand miles long; or a road as long as from Philadelphia to St. Louis, Mo., or from Boston nearly to Chicago. To remove them from the bakery in wagons, allowing 500 loaves for each, and take a load every half hour, to be thrown into the Delaware River, and continue this for ten hours a day during the entire year, would require 164 wagons to haul these loaves to the river in one year, or one wagon in 164 years.

What a thrill of horror would be excited in the breast of every sane citizen of Philadelphia if these 164 wagons should be seen going down Market Street to the Delaware, each having 500 4-pound loaves of bread to be thrown into the river ; and we feel safe in saying that not a single loaf would touch the water before he who would attempt to destroy so much food would be thrown after it. Yet, year after year, there is grain destroyed in the manufacture of intoxicating drinks equal to the amount of bread that those 164 wagons could haul, at two loads an hour, working ten hours a day for the whole year. If the six hundred million loaves of bread were annually destroyed by being cast into the rivers of the country, at the most and worst the bread would be lost, and that would be the end of it; the destruction of this bread would be a blessing to our people compared with the results that flow from the intoxicating drinks made from this wasted grain. The drink not only ruins our people financially, but undermines their virtues, blunts the sensibilities, effaces the memory, enfeebles the understanding, dethrones reason, and destroys life.

It cannot be denied that the grain is wasted in the process of malting, brewing, and distilling.

The food thus annually wasted would feed millions of our people. It is a sin and a crime to destroy food, though enough may still remain to feed the people. Every bushel of grain that is made into liquors enhances the price of what remains in the market; and dear bread always causes bad trade, for the more people have to pay for food, the less money they have for clothing and other comforts or luxuries. The results are the same, whether forty million bushels of grain perish in the fields by rain and mildew at harvest-time, or are subsequently destroyed in the breweries and distilleries. In both cases the price will be raised ; but in the latter case there is not only the destruction of the grain, but the destruction of the virtue of our people, the disinclination to engage in useful and productive labor to make up for the increase in the price of food, which is a twofold loss to the community.

In addition to these we have the immeasurable evils and burdens that flow directly and indirectly from the use of alcoholic drinks.

It is very clear, then, that if the grain was used for bread, instead of being destroyed in our breweries and distilleries, it would be vastly better for all classes of our people.

No nation can prosper long that practises such waste of food.

CHAPTER X.

THERE is no more difficult task than that of undertaking to find out the true cost of pauperism and crime in the United States. In truth, it may be said to be impossible from the poor and irregular system, or no system, of collecting facts and statistics in the public institutions of the country. There are only one or two States where any reliable statistics are collected. The figures of these States will be given, from which we may be able to proximate the condition of some of the others.

Though the figures presented are approximated, they are yet sufficiently near to convince every reflecting person that each of these evils is an immense pecuniary burden upon the industry of our people.

By viewing the vast resources of our country, given in the first chapter, we should be led to conclude that there must be general prosperity throughout our country, and that all our people would be well clothed and bountifully fed, and that no want or poverty could exist in all our favored land.

But on looking around us and viewing the actual condition of the masses of our people, we are forced to the very sad and painful conclusion that while our nation has been growing in wealth, and has year by year been extending its means, and increasing the appliances to produce more and greater wealth, large numbers of our people have been growing poorer and poorer, and that now tens of thousands are already in the midst of hardships and penury, and are either supported as paupers in our public institutions, or, what is still worse, both for them and society, as beggars and vagrants by private charity.

It will be seen, by a table hereafter to be given, from the census returns of 1870, that in the United States during the year ending June 30, 1870, there were 116,102 persons in the different poor-houses; that 76,737 received support June 1, 1870, at a cost of $10,930,429. And also that during the same period there were convicted 36,562 criminals, and that there were on that day 32,901 of this class of persons in the prisons of the United States. In the same year there were 143,115 licensed retail liquor-dealers in the United States. The census returns of paupers do not exhibit the full extent of pauperism and vagrancy.

Extreme poverty is not confined to those receiving regular or temporary relief from public institutions; for thousands of our laboring classes who never apply for public charity suffer untold hardships for want of the necessities and comforts of life.

TABLE XVI.

*This Table shows the Pauperism and Crime in the several States
from the Census Returns of 1870.*

States and Territories.	Populati'n, 1870.	Pauperism.		Crime.		No. licensed retail liquorshops,etc., 1870.
		No. Supported dur'g the year 1870.	Cost of annual support.	No. convicted in the year 1870.	No. in prison June 1, 1870.	
			Dollars.			
United States......	38,558,371	116,102	10,930,429	36,562	32,901	143.115
Alabama.	996,992	890	81,459	1,269	593	1,976
Arizona............	9,658	....	....	29	11	119
Arkansas	484,471	626	74,917	343	362	2,000
California..........	560,247	2,317	273,147	1,107	1,574	5,845
Colorado	39,864	73	11,422	32	19	371
Connecticut	537,454	1,728	189,918	450	430	3,353
Dakota.............	14,181	....	....	2	3	82
Delaware...........	125,015	556	41,266	145	66	368
Dist. Columbia.....	131,700	303	26 364	121	143	1,087
Florida	187,748	147	9 830	335	179	580
Georgia............	1,184,109	2,181	159,793	1,775	737	2,767
Idaho......	14,999	41	7,247	26	28	244
Illinois............	2,539,891	6,054	556,061	1,552	1,705	8,562
Indiana	1,680,687	4,657	408,521	1,374	907	4,444
Iowa	1,194.020	1,543	175,179	615	397	3.073
Kansas.............	364,399	361	46,475	151	329	1,117
Kentucky	1,321,011	2,059	160,717	603	1,067	4,761
Louisiana..........	726,915	590	53,300	1,559	845	4,414
Maine.............	626,915	4,619	367,000	431	371	843
Maryland	780,894	1,857	163,584	868	1 085	4,285
Massachusetts.. ...	1.457,351	8,086	1,121,604	1,593	2,526	5,039
Michigan...........	1,184,059	3,151	269,682	835	1,095	5,020
Minnesota..	439,706	684	66,107	214	129	1,930
Mississippi........	827,922	921	96,707	471	449	1.807
Missouri	1,721.295	2,424	191,171	1,503	1,623	5,888
Montana	20,595	104	17,065	24	16	449
Nebraska	122,993	93	11,161	53	69	635
Nevada	42 491	196	23,702	132	99	658
N. Hampshire......	318,300	2,636	235,126	182	267	1,161
New Jersey........	906,096	3,256	283,341	1,040	1,079	5,649
New Mexico.......	91,874	....	....	95	24	418
New York.........	4,382,759	26,152	2,661,385	5,473	4,704	21,318
North Carolina....	1.071,391	1,706	136.470	1,311	468	1,315
Ohio	2,062,260	6,383	566,280	2,560	1,405	11,769
Oregon............	90,923	133	24,800	80	104	738
Pennsylvania	3,521,951	15,872	1,256,024	3,327	3,231	13.015
Rhode Island.......	217,353	1 046	97,702	209	180	727
South Carolina.....	705,606	2,343	224.805	1,399	732	1,565
Tennessee.........	1,258,520	1,349	99,811	722	981	2,684
Texas.............	818,579	204	21,219	260	732	2,168
Utah..............	86,786	56	6,206	27	19	128
Vermont...........	330,551	2,008	178,628	139	193	540
Virginia...........	1,225.163	3,890	303.081	1,090	1,244	3,314
Washington	25,965	34	5.283	20	19	224
West Virginia......	442,041	1,102	80,628	155	191	543
Wisconsin	1,054 670	1,553	151,181	837	418	3,864
Wyoming.....	9,118	....	....	24	13	236

TABLE XVII.

General Statistics of the Town Paupers; showing the whole number fully supported, partially supported, etc.; the whole cost of all kinds of support and relief and other particulars of the Paupers of Massachusetts since 1854.||

Year.	Whole number, including vagrants of the towns' poor supported.	Whole number of paupers in and out of Alms-house.	Number of Alms-houses.	Average number supported in Alms-houses.	Average weekly cost in Alms-houses.	Whole No. of persons supported or relieved out of Alms-houses.	Number of insane poor supported.	Number of idiotic poor supported or relieved.	Total expenses.	Population.
					Dollars.				Dollars.	
1854	23,125	10,088	192	3,524	1 32.2	12,557	864	345	457,506 51	
1855	18,227	5,220	194	2,595	1 34	11,756	582	289	437,661 01	
1856	21,102	5,045	209	2,944 78/760	1 44	15,858	634	280	484,869 93	
1857	24,905	7,714	213	3,554 161/430	1 53	17,244	666	341	521,254 61	
1858	37,206	11,845	212	3,254	1 57	23,071	870	306	550,019 84	
1859	31,400	10,369	222	3,105½	1 47	21,954	816	326	522,312 93	
1860	34,314	7,787	219	3,290	1 51.4	14,023	852	293	545,245 46	
1861	52,847	9,374	219	3,385½	1 45.2	19,936	749	243	643,837 23	
1862	49,991	5,391	220	3,377	1 34.4	19,729	856	314	662,601 45	
1863	43,020	4,886	218	3,233	1 39.4	35,207	811	275	610,862 00	
1864	‡36,000	*5,000	218	2,866.24	1 70	†21,000	833	360	546,847 15	
1865	‡45,000	5,316	218	2,896.56	1 73	‡23,000	925	379	610,728 73	1,094,848
1866	52,628	5,715	222	2,984.37	1 98.5	24,335	974	380	746,159 68	1,094,848
1867*	57,251	5,862	223	2,960.51	2 15.2	26,918	1,124‡	436‡	758,360 46	1,094,848§
1868	91,157‡	5,706	224	3,010.22	2 37	29,648	1,207	469	832,501 65	1,094,848
1869	84,779‡	5,633	225	3,004.25	2 26.8	24,750	1,268	418	837,018 40	1,094,848
1870	64,870‡	5,553	225	2,753.44	2 55.6	25,203	1,320	427	854,609 56	1,457,353

Compare the Statistics of 1867 with the subsequent year; for in 1867 the Prohibitory Law was more rigidly enforced than any year since.

* Approximated.
† Including lodgers at the Boston Station-house, 25,000 annually.
‡ Excluding lodgers at the Boston Station-house, 25,000 annually.
§ The year when the Prohibitory Law was partially enforced, and the only year until 1873.
|| Report of State Charities, 1870, p. 432.

TABLE XVIII.

General Statistics of Town Paupers for 1867, showing the whole number of Paupers supported September 30, 1867, the number partially supported, and cost of all kinds of support, etc.*

Counties, 1866-7.	Population, 1865.	Whole No. within and without the Alms house.	No. sup-ported Sept. 30, 1867.	Paupers Insane.		Paupers receiving partial support.	Cost of Paupers.			Total cost of Support and Relief.
				Whole No. sup-ported.	No. Sept. 30, 1867.		At the Alms-house.	Out of Alms-houses.	Total full Support.	
							Dollars.	Dollars.	Dollars.	Dollars.
Barnstable......	34,610	171	148	29	25	580	15,510 85	2,008 66	15,519 51	30,505 41
Berkshire......	56,944	152	131	18	15	463	1,929 23	11,844 55	13,773 78	20,132 15
Bristol.........	89,395	568	353	57	46	3,132	37,677 05	8,580 49	46,257 54	77,308 45
Dukes..........	4,200	44	42	8	8	75	1,837 62	3,109 91	4,947 43	6,525 54
Essex..........	171,034	874	608	147	129	4,922	62,603 28	14,186 70	76,789 98	139,267 71
Franklin.......	31,340	183	140	40	35	251	7,101 10	8,197 91	15,299 01	19,598 95
Hampden.......	64,570	270	108	47	41	879	8,197 00	13,829 82	22,326 82	30,092 66
Hampshire.....	39,269	164	136	35	24	265	4,439 71	12,245 99	16,685 10	21,105 75
Middlesex......	220,384	932	658	117	97	3,925	60,110 30	10,704 49	71,814 79	100,982 27
Nantucket......	7,748	103	72	5	5	280	5,413 34	1,969 36	7,382 70	11,538 71
Norfolk........	116,306	448	307	82	75	1,803	31,962 14	16,428 59	48,390 73	71,036 68
Plymouth.......	63,107	305	243	69	57	764	25,185 90	5,012 88	30,198 78	45,942 58
Suffolk........	208,212	883	341	169	134	5,749	33,431 89	37,738 23	71,170 12	106,525 88
Worcester......	102,912	765	560	125	106	2,926	40,463 03	16,367 02	56,830 05	77,737 74
Totals......	1,267,031	5,862	3,907	948	795	26,014	335,161 74	162,224 60	497,386 34	758,360 41

* Report of State Charities, 1868, page 298.

TABLE XIX.

*Showing the average number of inmates, average weekly cost, the cost of medicine, etc., current expenses, etc., etc., of the ten institutions of Massachusetts named for the year 1867.**

Institutions.	Average No. of inmates.	Average weekly cost.	Medicines and medical supplies.	Provisions and supplies.	Salaries, wages, and labor.	Total expenses.
Worcester Hospital	389	$4 30	$1,179 22	$33,534 90	$21,027 18	$86,930 88
Taunton Hospital	379	3 60	664 26	31,398 75	13,333 26	70,937 83
Northampton Hospital	401.03	3 80	692 26	34,005 91	15,273 85	90,649 76
Rainsford Hospital	1	....	35 08	958 15	2,630 44	4,953 18
Tewksbury Almshouse	757	1 77	504 01	34,847 48	10,426 83	69,583 82
Monson Almshouse	628.5	1 90.8	362 54	29,075 78	10,755 37	67,648 59
Bridgewater Almshouse	331	2 00	437 58	17,861 08	6,488 84	52,222 41
Westborough School	326	3 08	65 89	22,790 89	13,747 36	60,653 73
Lancaster School	141	3 38	108 00	7,202 20	9,642 61	25,531 11
School-Ships	285	3 07	322 91	23,865 92	15,081 48	57,035 98
Totals	3,638 63/100	....	$4,371 75	$235,541 06	$119,007 22	$556,147 29

* Report State Charities, Mass., 1866, pp. 69, 70.

TABLE XX.

Showing the number of paupers in the towns' almshouses by counties, Massachusetts, with other statistics, in 1867.*

	Whole No. fully supported.	Whole No. relieved and partially supported.	Whole No. of vagrants, including those sent to State almshouses.	Average No. at almshouse.	Weekly cost, average.	Expense at almshouses.	Expense out almshouses.	Total expenses.
Barnstable	163	576	13	139.36	$1 86.4	$13,510 85	$15,956 22	$29,467 67
Berkshire	30	215	229	18.5		1,929 23	1,922 12	3,851 35
Bristol	568	3,132	1,849	337.31	2 15	37,077 05	39,631 38	77,368 43
Dukes	16	65	2	11	3 21.2	1,837 52	1,212 24	3,049 76
Essex	823	4,774	2,028	512.87	2 31	61,609 95	68,538 72	130,148 67
Franklin	109	160	85	79.65	1 72.4	7,101 10	3,997 15	11,098 25
Hampden	159	598	2,376	82.58	1 98	8,497 00	5,453 17	13,950 17
Hampshire	68	161	341	47.41	1 80.4	4,439 11	3,707 47	8,146 58
Middlesex	889	3,559	5,219	574.86	2 00.4	59,910 30	33,739 93	93,650 23
Nantucket	103	280	1	56.56	1 64.6	5,413 34	6,125 37	11,538 71
Norfolk	428	1,750	3,757	239.60	2 56.5	31,962 14	29,557 14	61,519 28
Plymouth	282	723	476	200.18	2 29.3	23,925 79	17,743 42	41,669 21
Suffolk	872	5,398	1,237	176	3 65.2	33,431 89	68,683 99	102,115 88
Worcester	724	2,833	4,424	484.63	1 62 1	40,463 03	31,414 50	71,877 53
Totals	5,238	24,224	22,937	2,060.51	$2 15.2	$331,708 30	$327,082 82	$639,391 32
Townships in the above counties having no almshouses—totals	578	1,432	2,550					90,924 65
Total for State	5,816	25,656	25,487					$660,316 47

* Report State Charities of Massachusetts, 1868, pages 260, 261.

PAUPERISM AND CRIME IN THE YEAR 1868 IN
PENNSYLVANIA.

The Citizens' Association of Pennsylvania, charter-
tered by the Legislature "to report on the dependent
and criminal population of the State," in their report
to the Legislature dated February 1, 1868, gave the
following facts:

"The paupers in poorhouses and chargeable to
counties numbered 14,988, or *one* in 246 of the
population. Cost of maintaining them at 29 cents*
per day each, or $106 60 per year, amounts to $1,597,-
720, or $2 67 for each voter in the State.

"The percentage of the public poor who are helpless
from age, disease, or other infirmity is about .45,
leaving .55 who are able to employ themselves in
some occupation that may in part remunerate the
counties for their support.

"Relief given to deserving poor, . . . or out-
door relief, amounts to $190,376 56, or 32 cents to each
voter.

"The number of the second class of poor, denomi-
nated vagrants, cannot well be ascertained, but from
returns in hand the number of meals furnished to
such at the poorhouses is estimated at 361,000, which,
at 15 cents per meal, would amount to $54,150, or 9
cents to each voter.

"The number of nights' lodgings furnished to
travelling poor is 119,096. Add to this the lodgings

* The average of the Philadelphia Almshouse, which is lower than any
other in the State.

in station-houses in Philadelphia, 46,250, and we have a total of 165,346 nights' lodgings furnished to vagrants."

Two-thirds of the above pauperism and three-fourths of vagrancy are justly attributed to intemperance or the use of intoxicating drinks.

From this report of the Citizens' Association* we find that one-third of the insane, deaf-mutes, blind, and feeble-minded are attributed to intemperance; and also that two-thirds of the friendless children and the inmates of the houses of refuge, or 1,154 of these dependents in the State, are from the same cause.

They also say: "The estimated population of county jails is 8,447; of penitentiaries, 669, or one in 402 of the population. The average cost for the maintenance of these prisoners is 44 cents each per day—a total per day of $4,011 04, or $1,464,029 60 per year, or to each voter in the State $2 45.

"Causes.—It will not be doubted that two-thirds of the pauperism and crime of the State are justly attributed to intemperance, and it is stated by authorities that one-third of the dependent classes—as insane, feeble-minded, etc.—are to be traced to the same cause. If we apply this rule to the figures before us, we have the aggregate cost of maintaining paupers and criminals whose condition is due to intemperance $2,204,244 per year, and the aggregate cost of maintaining insane, idiotic, and other depen-

---

* Report Citizens' Association, Pennsylvania, 1868, page 9.

dent persons from the same cause $55,066 66,—a total cost of $2,259,910 66.

"These are startling facts which deserve candid thought, and should be taken into account by legislators and all persons who have an interest in public morals and in the economy of our State affairs. . . . Ought we not to ask: If we have done so much for the *support* of pauperism and crime, what have we neglected to do for the *arrest* of these evils? It seems to us that if both sides of this question are fairly examined, our sins of omission will rise up against us with fearful condemnation. . . . .

"The victims of strong drink, however, come in hosts more numerous than all the rest together, and with hopes blasted, self-respect gone, and the story of domestic sorrow and grief bearing upon the heart, point to the path of ruin that is before them, and ask for help.

"Thirty thousand people in Pennsylvania are in this condition, and come to ask you (the legislators of Pennsylvania) with these pictures of pauperism, dependence, and crime, asking that they may have a share of your sympathy—not that they may be abased and imprisoned as criminals, nor yet be humiliated as paupers, but that they may have such help as will enable them to be men again and do their portion for the public good."

ALMSHOUSES, PENNSYLVANIA, 1871.

The amount expended in Pennsylvania in 1870 for relief and maintenance in Almshouse and out-door Paupers, with cost per capita, the number of Intemperate, Insane, Idiots, and Vagrants, from the counties and places named.

Almshouses.	Expended for Maintenance	Out-door Relief Expenses.	Total Expenses, Indoor and out.	Cost per capita per Week.	No. Supported and Relieved.	No. Intemperate.	No. Insane.	No. of Idiots.	No. Vagrants Relieved.
	Dollars.	Dollars.	Dollars.	Dollars.					
Alleghany City Poorhouse,	19,937 53	3,985 36	23,922 89	1 07	256	192	32	8	424
Beaver County Almshouse,	3,119 49	300 00	3,419 49	1 12	70		10	9	
Bedford County Almshouse,	12,791 17	1,631 98	14,223 15	1 15	133	41	5	21	
Berks County Almshouse,	28,650 35	2,043 77	30,694 12	1 86	624		37	6	
Blair County Almshouse,	8,000 00		8,000 00	3 00	148	90	2	2	
Bucks County Almshouse,	22,294 65	1,941 00	24,235 00	1 64	417				1,460
Cambria County Almshouse,	10,375 30	2,421 61	12,796 91	1 50	111		18	2	932
Middle Coalfield Poorhouse, Carbon Co.,	10,266 78	2,031 90	12,298 27	2 00	153	40	8	1	30
Chester County Almshouse,	14,805 41	1,106 90	15,912 31	1 11					
Lockhaven Poorhouse, Clinton Co.,	1,415 00		1,415 00	1 84	13				
Centralia Poorhouse, Columbia Co.,	2,119 47	1,781 34	3,900 81	2 00	58	1	16	5	
Poor District Bloom Poorhouse, Columbia Co.,	159 75		159 75		18	1		1	
Crawford County Almshouse,	11,159 77	2,723 47	13,883 24	1 50	171		8		6,932
Cumberland County Almshouse,	19,000 00	2,500 00	21,500 00	2 00	274		29	4	5,188
Dauphin County Almshouse,	33,458 00		33,458 00		409	162	30	11	
Delaware County Almshouse,	15,759 54	2,836 71	18,506 25	1 76	367		13	18	
Erie County Almshouse,	14,422 07		14,422 07		163				
Fayette County Almshouse,	8,801 76	1,698 02	10,609 78	1 26	78				

ALMSHOUSES, PENNSYLVANIA, 1871—Continued.

Almshouses.	Expended for Maintenance.	Out-door Relief Expenses.	Total Expenses, indoor and out.	Cost per capita per Week.	No. Supported and Relieved.	No. Intemperate.	No. Insane.	No. of Idiots.	No. Warrants Relieved.
	Dollars.	Dollars.	Dollars.	Dollars.					
Green County Almshouse,	2,511 35		2,511 35	1 20	64	3	5	10	
Huntingdon County Almshouse,	5,757 11		5,757 11	1 50	75	12	7		8,102
Lancaster County Almshouse,	32,784 19	9,763 00	42,547 19	1 25	607	450	77	2	5,452
Lebanon County Almshouse,	21,062 00	2,921 47	23,983 47	2 00	164		11		
Lehigh County Almshouse,	23,244 97	2,039 25	25,284 22		474	54	4	2	
Central Poorhouse, Luzerne Co.,	5,598 83	985 99	6,584 82	1 50	58	13		1	
Lackawanna Poorhouse, Luzerne Co.,	1,231 75	243 00	1,474 75	2 75	33	50		3	
Carbondale City Poorhouse,	802 62		802 62	2 00	63	57	10	1	
Mercer County Almshouse,	5,470 21		5,470 21		115	8	7		25
Mifflin County Almshouse,	5,380 00		5,380 00	1 09	67			3	
Montgomery County Almshouse,	15,899 54	5,616 37	21,515 91	1 75		13	44		
Northampton County Almshouse,	11,599 15		11,599 15	2 16	489		26	1	
Perry County Almshouse,	5,474 86	825 50	6,300 36	1 78	72				
Philadelphia County Almshouse,	287,456 50	86,922 75	374,379 25	1 82	9,980	540		1	4,512
Washington County Almshouse,	1,832 73	1,453 36	3,286 09	1 29	198	18	3		
Philadelphia and Germantown Poorhouse,	10,802 05	1,817 32	12,619 37	2 00	98	7	1	3	
Philadelphia, Oxford, and Lower Dublin Poorhouse,	13,634 78	2,388 18	16,022 96	2 00	94	60	1	10	2,073
Schuylkill County Almshouse,	37,581 21	5,375 91	42,957 12		651		55	3	
Susquehanna, Montrose, and Bridgewater Almshouse,	1,364 75	171 30	1,536 05	1 66	13		4	1	
Tioga County Almshouse,	4,696 07	811 31	5,507 38	2 50	96	3	5	4	50
Warren County, Rouse Hospital,	6,400 00	300 00	6,700 00	2 00	54			1	
Wayne County Almshouse,	2,800 00		2,800 00	2 00	10				
Westmoreland County Almshouse,	16,191 17	5,683 72	21,874 72		264	29	43	38	
York County Almshouse,	35,875 19	1,805 51	37,680 70	2 50	356		8	3	1,439
Totals,	791,987 07	156,325 42	948,312 49	1 75	17,571	1,955	523	180	36,619

The Board of State Charities sent interrogatories to nearly 700 districts ; 212 only replied,* from which we find that the cost of pauperism in those districts was $68,538 92 for 760 paupers, of which 64 are reported to be intemperate, 71 insane, and 35 idiotic. These figures are very far from being correct, for the State Board of Charities report says : "The accompanying tables of almshouses and township poor . . . illustrate the great necessity of further legislation, requiring a uniform system of statistical records to be kept in these institutions. Some are not able to give the number of persons relieved during the year, the average number admitted, or the weekly cost of support. . . . In many no distinction is made as regards sex, color, nativity, etc., etc. While a majority of stewards allege intemperance to be the cause of pauperism, yet there are no regular records kept of the number of intemperate persons receiving support ; and when asked, How many of those supported or relieved were intemperate? answer frequently, 'We have no record upon the subject.' Hence the difficulty of arriving at the real cost of pauperism."

The following table shows the number of paupers, the amount expended, number intemperate, insane, and idiots, and children under sixteen years, in the county almshouses and districts of Pennsylvania :

* Board of Public Charities Report, Pa., 1871, p. 94.

Names.	No. Paupers relieved.	Amount Expended	No. Intemperate.	No. Insane.	No. Idiotic.	Under 16 years.
		Dollars.				
County Almshouses. .	17,571	948,312 49	1,955	523	180	1,075
* Township Poor.....	760	68,538 92	64	71	35	300
† Totals.......	18,331	1,016,851 41	2,019	594	215	1,375

From the Report of the State Board of Charities and Reform of Wisconsin for 1871 we learn that for 1870 the whole number of persons receiving relief of towns was 3,800; the total for the State is estimated at about 5,000. The amount expended for the relief of poor was:

Cost of county poorhouses, ...	$70,553 09
Relief of poor not in poorhouses from county treasuries, ...	69,307 78
Amount paid from town treasuries, . . .	113,004 57
Total, ...	$252,865 44

The United States census gives the number of persons supported in Wisconsin during the year 1870 as 1,553, while the reports to the Board of Charities, etc., are 3,792; and there is good evidence, as the Board says, to believe that some ten or twelve hundred more were relieved that were not reported to them. The census returns give the cost of pauperism as $151,181, while the returns to the Board show a cost for the year ending a few months later than that covered by the census to have been $252,864 44; licensed places to sell liquors, 2,613; number of places without licenses selling liquor, 414.

There is also in every State a large class of persons

* Board of Public Charities Report, Pa., pp. 98–104.
† Ibid., pp. 106–9.

who mainly depend upon their relatives and friends for support.

We feel safe in saying that twice or even thrice as many persons are in the condition of paupers as are reported to have been relieved in our alms-houses.

The returns of vagrancy are even more imperfect than those of pauperism; but every one must be painfully aware that in every part of our country there is a very large class of persons who have no fixed place of abode, but are moving about from place to place, and obtain a living by begging and stealing, or by some of the many impositions practised upon the public. This class of persons is a worse burden upon society than our actual paupers; for they are not only supported by the public, but they carry an atmosphere of demoralization wherever they go; 1,408 of this class of persons were arrested by the police of Philadelphia in 1873, being nearly 700 more than the previous year.

To survey the extent and magnitude of our resources; the power for greater production and development, and the accumulation of still greater wealth; the diversity of our climate; our numerous majestic rivers, which peculiarly fit us to become a great manufacturing, agricultural, and commercial nation, we can but conclude that something must be radically wrong by which so much vagrancy and pauperism exist amid such resources and natural advantages. But when we consider the results flowing from our numerous drinking-places, the demoralizing social habits caused thereby, and sum up the money that is squandered for intoxicating drinks, we can no longer wonder at the poverty, misery, and pauperism which exist; for there is, and ever will and ever must be, a never-failing relation and connection between the facilities for obtaining intoxicating drinks and vagrancy and pauperism. As these facilities and consequent drunkenness are increased in any town, city, or State, in that ratio is pauperism augmented. Every cent taken from the pockets of our laboring classes for liquors is taken directly from the means of procuring the necessaries of life, and is a total loss to them.

It is this squandering of money for strong drink by our laborers, mechanics, artisans, etc., which places or keeps them in that condition wherein, if by sickness, accident, depression of business, or other cause they are unable to follow their avocations, they either greatly suffer or become burdens on the charities of the public. A laboring man need not become a drunkard to impoverish himself and family. To drink two or three glasses a day is sufficient to produce want or a lack of many comforts of life. Then, in addition to the loss of the money so spent, the continual though moderate use of the liquor so poisons and undermines the drinker's constitution that very often before he has arrived at the meridian of life he is a worn-out old man, a dependent, and his family a burden on the sober, healthy, and industrious. Not less than 130,000 of the widows and orphans, annually left in our country, are left such by the liquor-

drinkers. From two-thirds to four-fifths of the inmates of our poorhouses are there by drink.

If the individual and the family history could be
ascertained, it would be found that not less than nine-
tenths were brought directly or indirectly to the condition of inmates of the almshouse either by the intemperance of themselves or others.

It is true that sometimes, by commercial depression,
misfortune, sickness, or other causes, persons may be
brought to poverty and distress ; but for the use of
intoxicating drinks, few indeed would be the cases
that would need to be sent to the poorhouse. Every
liquor-shop is a moral plague-spot and hot-bed of
disease and destitution.

This is not only seen to be the case from the returns
of our pauper institutions, but in the mendicity that
exists throughout the land. The greater part of the
beggary is created and perpetuated by the traffic in
strong drinks. The expense to our people collectively
of the beggary so increased in our country is little if
any less than the pauperism in our public institutions. Our working classes will inevitably be kept
poor and dependent so long as the temptations of the
liquor-traffic exist ; for, as a general rule, at least all
their surplus earnings are spent for drink. Why is
it that the houses of the liquor-sellers are well furnished and often owned by themselves, while the
homes of those who patronize them are destitute not
only of the comforts but even the necessaries of life ?
It is because of the self-imposed taxes they burden
themselves with for drink, and deprive themselves

and families of the pleasures and the comforts of home-life for the degrading, demoralizing, and momentary enjoyments and delusions of drink. The greater number of our working-men expend for liquor sums of money which, if saved for a few years, would purchase handsome and well-furnished homes or make provision for sickness, accident, or old age. For example, let us make a simple calculation of the sums of money spent for liquor by a very moderate drinker. We will suppose that a young man commences at the age of 20 years to drink, and that from 20 to 23 he drinks but one glass of beer a day, worth 5 cents a glass; at 23 he will have spent $54 75; from 23 to 25, two glasses a day, he will have spent $73; from 25 to 30, three glasses a day, $273 75; from 30 to 35, four glasses a day, $365; from 35 to 40, five glasses a day, $456 25.

Thus, a young man commencing at the age of 20 to drink in the strictest moderation will have spent at the age of 40 for beer, which did him not one particle of good, but more or less injury, the sum of $1,222 75. Now, if another young man commences at 20, and, instead of spending the money named for beer each year, should put it out at 6 per cent. interest, without any other savings but this beer-money he would be worth, at the age of 40 years, $1,955, having saved his money, his character, his health, and perhaps his soul. "And what will a man give in exchange for his soul?"

Young men, ponder well the above. Is not the allowance we have given less than the quantity of liquor drunk by the average moderate drinkers in the course of 20 years? Let the most temperate drinkers reflect whether they do not annually spend more for strong drinks than the sum named. Is it not foolish, ay, sinful and criminal, to spend hard-earned money for that which is not only of no benefit, but injurious?

A certain man was in the habit of saying when he drank a glass of liquor, "Here goes a peck of potatoes." But the man who has drunk only moderately for twenty years may say with truth, "I have swallowed a three-story house and lot, chimney and all."

Our industrious classes are not only injured by the money they spend for intoxicating drinks, but by the time lost from drinking, which, even in the case of moderate drinkers, is often more in value than the money expended.

Again, the use of intoxicating drinks tends to create spendthrift and improvident habits, making the poor man indifferent or content in his poverty, lessening his self-respect, destroying his laudable ambition, which prevents him from trying to better his circumstances. This is clearly exhibited in almost every family where strong drinks are used to any great extent. Total abstinence has an opposite tendency; for as soon as men who have been in the habit of drinking give up their cups they become more industrious and ambitious, provide better for the wants of their families, and try to get along better; and many, in a short time, commence to accumulate property and "provide for a rainy day."

Nothing injures our working classes so much as

drinking. The man who from month to month spends one-third or one-half of his earnings in drink is in no condition to resist the reduction of wages or unjust exactions of his employer. He is compelled to submit and accept any terms or wages offered, in order that himself and family may not be deprived of the necessaries of life. Drinking injures the sober, industrious mechanic, artisan, and laborer, as well as the dissipated; for they, too, are often compelled to accept a reduction of wages, as the number of dissipated workmen is so large at the present time that a sufficient number can always be found who must have work on any terms, or starve; hence the sober mechanic has to take less wages because the drinker is compelled to do so. Now, if the artisans, mechanics, and operatives, when trade is good, would save but half the money they now expend for liquor, there would be no need of strikes; for the workers would generally be in a pecuniary condition to resist any very unjust exaction of their employers.

For employers generally need the labor of their workmen as much as they need money; for it is out of the profits of their labor that the employers make their capital.

Let our laboring men take care of their money, and not expend it for liquor, and they will soon be as independent of their employers as the latter are of them; for, in a normal condition, capital and la-bor are equally dependent one on the other.

Again, if the industrious classes would save the money now spent for liquor, when their employers refuse or are unable to pay them remunerative wages they would be able with their savings to embark in other fields of productive labor which would not only benefit themselves, but relieve the labor market of a surplus of laborers. If, for instance, the 12,505,923 persons engaged in agriculture, manufactures, and mechanical and mining industries would deposit in our savings institutions annually one-fifth of the money now expended for liquor, or $140,064,276; or if the 2,053,996 persons engaged in manufactures, mechanical and mining industries, or about 410,799 families, would deposit in the savings-banks the $81 per year that is spent by each family in the United States for liquor, the accumulation of this sum in a few years, with the interest accruing, would be a very handsome reserve fund for each family to fall back upon in time of need.

With these savings at their command they would not be compelled, when business is a little dull, to work upon any terms their employers might offer.

Though strikes never benefit, we believe, any one, employer or employed, yet if the working-classes will abstain from liquors and save the money expended for them, in a few years, they might then strike with some show of success. But the strikes that take place now result generally to the injury of the working classes; for in the majority of cases, after losing weeks or months of labor, they are obliged to go to work without gaining what they struck for, because they had spent the money for drink that would have supported or enabled them to go into some other employment or become employers themselves. Hence, in any and every way we may look at this subject, it is to the interest of the working classes to adopt the principle of total abstinence; and the first strike should be against spending their money for drink and keeping the drink-sellers in idleness. By doing this they are sure to gain in money, health, and happiness.

Total abstinence will not only benefit the employed, but the employer. All other things being equal, the sober workman who totally abstains from all kinds of liquors is to be preferred to one who drinks. The non-drinking mechanic or artisan is generally able to do more and better work with greater ease to himself than the drinker. This is now certain; hence it is a loss for employers to have drunken hands, or even those who use strong drinks. Again, the non-abstainer will often neglect his work to spend his time in drinking. True, the employer does not pay his hands when they are not at work. The employer, when he engages a man, needs his work, and expects to profit by it; but when he spends his time in drinking, the employer not only loses the profit on the work he could have done, but his business is neglected, and often, as business is now carried on, other men may be kept waiting for the work he should have done. In such cases the employer not only loses the work of the drinker, but also that of the non-drinker, by drunkenness. If it is profitable to employ hands at all, it is certainly to his benefit to have sober

workmen upon whom he can depend; and it is just
as surely a loss to have men who drink. This was
well understood by Mr. Bokewell, of Manchester,
England, who offered to give a shilling a week extra
to every one of his workmen who should become a
worthy and consistent member of a total-abstinence
society.

It is strange that manufacturers and master-me-
chanics have not ere this become more fully awakened
to the loss they sustain by the drinking customs of
the country, not only by checking the development
of their industries, but by the loss they sustain from
the drunkenness and idleness of their employees.
Let us, to illustrate, suppose that Mr. A. has a ma-
chine-shop or factory, fitted up with machinery,
each part depending upon another. The success of
his business depends upon the skill and industry of
his workmen. He contracts to produce in a given
time a certain amount of the products of his business. To do this will require the steady and uniform
labor of one hundred hands to produce the manufactured articles by the time named. But, instead
of all these hands working regularly, there are eight
or ten hands every week or few days who lose their
time or neglect their work, either to drink or from
inability to work from the effects of drinking. The
consequence is that some portion of the machinery
is standing idle; and in order that the whole estab-
lishment shall not stop, he is obliged to keep his
engine going at a loss of fuel to turn a part of his
machinery, and the result is that the work will not

be done, unless he employs additional hands, or runs his machinery longer hours, and incurs the loss of light, fuel, and wear and tear of machinery. Thus will he be a great loser by the intemperance of his workmen, besides the trouble of mind and perplexity that will be experienced to have the contract completed in time. The same will apply to men in every business who are under the necessity of employing help. This is another of the great drawbacks upon industry. Hence there is no question that is agitating this country that so materially affects the interests of manufacturers, merchants, and tradesmen in every department as the right solution of the question arising from the drinking habits of the people of this and every civilized nation.

CHAPTER XI.

THE relations of the use of intoxicating drinks to crime is a subject well worth the serious consideration of statesmen and people, and is one which appeals to the sympathy and reflection of every lover of the human race. All must deplore the direful and demoralizing effects of the liquor-traffic upon our citizens, and particularly so when they consider the immense cost directly and indirectly caused by it.

In whatever direction we look, in every State and Territory of the United States, and in every portion of the civilized world, the terrible results of the use of, and traffic in, alcoholic drinks have been felt, and to which may be traced most of the crime, misery, and the disturbance of the public peace. This cause more than all others fills our jails, poorhouses, penitentiaries, and lunatic asylums, and does more to frustrate the efforts of Christians and philanthropists than all else combined.

In an article prepared by A. S. Fisk, A.M., entitled "The Relations of Education to Crime in New England, and the Facilities for Education in her Penal

Institutions," and published in the report of the United States Commissioner of Education for the year 1871, page 549, we find the following:

"The fourth fact is that from 80 to 90 per cent. of our criminals connect their courses of crime with intemperance. Of the 14,315 inmates of the Massachusetts prisons, 12,396 are reported to have been intemperate, or 84 per cent."

"At the Deer Island House of Industry (Boston), not included in the above figures, of 3,514 committals, 3,097, or 88 per cent., were for drunkenness; fifty-four more as idle and disorderly, which commonly means under the influence of drink; seventy-seven for assault and battery, which means the same thing; and forty-eight as common night-walkers, every one of whom is also a common drinker. We have, therefore, of this prison a full 93 per cent. whose confinement is connected with the use of drink; and this may be taken as a not exaggerated sample of many municipal prisons. In the New Hampshire State Prison sixty-five out of ninety-one admit themselves to have been intemperate. Reports were asked from every State, county, and municipal prison in Connecticut in the spring of 1871 in reference to the statistics of drinking habits among the inmates, and it was found that more than 90 per cent. had been in habits of drink by their own admission.

The warden of the Rhode Island State Prison, and county jailer, estimates 90 per cent. of the residents of his cells as drinkers.

From Vermont and Maine no reports have been secured ; but they would not, if their prisoners were all interrogated, bring the estimate below 80 per cent.

It will still be remembered that these figures do not cover the mere temporary arrests for drunkenness, disorder, etc., nor the facts of the municipal place of detention, where the percentage of drunken criminals will be most striking.

There is no enormity or crime to which persons, no matter how well disposed and gentle at other times, may not be impelled when under the influence of drink.

Husbands and fathers are not only caused to neglect wives and families, but to inflict upon them the most revolting cruelties. The affections in families are blunted and obliterated; children are neg-- lected and left without clothing, food, or education, and often forced into crime by their parents to pro- cure money for them to spend in drink, or they are abandoned and left to shift for themselves, and under the guidance of wicked associates are urged to com- mit crime to eke out a shiftless existence.

There can be no doubt in the minds of any who have examined the subject in the least but that the liquor- traffic is the main source and prolific cause of the criminality that is steadily increasing from year to year, and which consequently necessitates the increase and enlargement of prisons and police officers. All of which has again and again been fully and clearly established by the testimony of judges, grand-juries, police magistrates, chaplains, governors, and inspec- tors of prisons. They have repeatedly testified that

frauds, embezzlements, theft, the prostitution of our
young women, robberies, burglaries, and murders, are
produced mainly by the brutalizing and depraving influences of strong drinks.   More than three-fourths of
the inmates of prisons attribute their fall to the use of
intoxicating drinks.   Of the 39 cases of murder and
121 cases of assault to murder in the city of Philadelphia in 1868, in almost every case it may be safely
said that the murderer was intoxicated when the
deed was committed.   These bloody deeds were clearly traceable to the liquid poison that maddens the
brain, depriving of reason, and leading to the commis-
sion of acts of blood and violence at the thoughts of
which, when sober and clothed in their right minds,
the perpetrators' souls would revolt.   They would
say with one of old, "Is thy servant a dog that he
should do this great thing?"   For all these evils flowing from the liquor-traffic not only do heavy and
fearful responsibilities rest upon the liquor-sellers,
who entice, by various means, men and women to
enter their places and indulge in strong drink, but a
terrible responsibility is also laid at the door of the
law-makers, citizen voters, and every one who does
not exert all his influence, political, social, and religious against legalizing such traffic.

Reader, do you doubt that intoxicating drink produces the crimes charged against it ? If you do, examine well the following figures and facts.

The Brewers' Congress and the Liquor-Dealers'
Associations boast of the great revenue they pay for
the privilege of selling liquors.   The amount paid for

tavern licenses in Pennsylvania in 1867 was $279,532; for beer licenses, $40,482—making a total of $320,015. Of this sum $162,746 was paid in Philadelphia. During that year, of 36,333 persons arrested in the city of Philadelphia, 13,930 were committed to prison for drunkenness who were not able to pay their fines, etc., but were incarcerated at the expense of the public. There were committed to the Philadelphia County Prison, from the 1st of January, 1868, to January 1, 1869, for drunkenness, vagrancy, disorderly conduct, and breaches of the peace, 9,220. In the year 1867, as already seen, Pennsylvania paid for criminal and pauper expenses caused directly by liquor-drinking, $2,259,910, or an average of $5 80 for each voter in the State. The same year Philadelphia paupers and criminals cost $1,500,000, or $11 for each voter. What did Philadelphia receive in the way of revenue from license towards paying this million and a half of dollars? Nothing. The money paid for licenses went into the State treasury. The State received $317,742 75 for licenses to sell liquor, and paid for pauperism and crime caused by the use of strong drinks $2,259,910; or, in other words, the State from licenses received 14 cents, and spent one dollar for crime and pauperism. Truly, the State paid dear for its whistle. But, to be more specific in our charges against the liquor-trade, we will present a few facts from official records.

The report of the Board of State Charities of Pennsylvania for 1871, on page 89, says: "The most prolific source of disease, poverty, and crime, observ-

ing men will acknowledge, is intemperance. In our
hospitals, as well as in our almshouses and prisons,
a large portion of the inmates have reached the refuge
in which they are found by the way of habitual intoxication." . . . "Intemperance, the great scourge
of society, is, as every one knows, a social vice. Few
inebriates begin their downward career by purchasing
the stimulant in quantity, and taking it home to use
at pleasure or convenience. The habit of its use is
contracted in some public place where like companions meet, and where the exhilaration which
strong drink produces may expand itself into bois-
terous mirth."

"The policy of giving licenses to certain parties to
open taverns, where intoxicating drinks may be par-
taken of, and gatherings may be accommodated for
their indulgence, is now in vogue." "The imposts
exacted for these licenses are a source of considerable
revenue." . . . On page 90 the report says: "It
would be difficult to name any practical good which
results from this system (of licensing liquor-shops),
unless it be that it furnishes a certain amount of revenue. Should these wages of iniquity be put into the
treasury ? They are the price of blood, and, in their
aggregate, would be inadequate to buy fields enough to
bury the multitudes who are the victims of the dreadful traffic for whose profits they sell the people's sanction." "And what economist can fail to discern,
without any elaborate calculation, that the State is
impoverished by the whole transaction ? There is
received into the public coffers a small tribute from

every man who cares to secure the common authority for the prosecution of this pernicious trade, and the consequence is that there is lost from the commonwealth the productive labor of thousands who waste, in the licensed haunts of intemperance, both the ability to add to her wealth and the accumulations of former thrift."

INTEMPERANCE AND CRIME—PHILADELPHIA, PA.

To form an idea of the amount of crime in Philadelphia we give the following:

Table showing the number of prisoners and cost for Philadelphia county prisons for ten years, from 1861 to 1871, inclusive:

In the year 1861 there were 16,201 prisoners, costing $50,643 59
" 1862 " 14,646 " " 50,745 25
" 1863 " 17,219 " " 50,225 95
" 1864 " 14,067 " " 58,737 51
" 1865 " 16,142 " " 69,252 51
" 1866 " 19,468 " " 103,111 13
" 1867 " 18,575 " " 95,276 60
" 1868 " 17,620 " " 104,631 63
" 1869 " 18,305 " " 105,625 12
" 1870 " 15,288 " " 102,630 03
" 1871 " 13,171 " " 163,807 55

The total for ten years was 180,501 prisoners, costing $894,736 92

Table showing the whole number of prisoners, before and after trial, confined in the County Prison of Philadelphia in 1871:

	Males.	Females.	Total.	Moderate drinkers.	Temperate.	Intemperate.
Prisoners received for trial..	4,423	1,105	5,528	....	....	....
Vagrants................	649	410	1,059	....	....	....
Disorderly and breach of peace	1,657	664	2,321	....	....	....
Intoxication............	2,721	963	3,684	....	....	....
Sentenced not to hard labor.	80	6	86	....	....	....
Sentenced to hard labor.....	306	47	400	135	157	108
Sent to Eastern Penitentiary.	71	1	72	....	....	....
Sent to House of Refuge....	20	1	21	....	....	....
Total...............	9,974	3,197	13,171	....	....	....

By the preceding table 13,171 were sent to the county prison of Philadelphia. After deducting the 5,528 sent for trial, there remained 7,643, of which number 3,684 were committed for intoxication; there were 2,321 cases of disorderly conduct and breach of the peace, and 1,059 vagrants.

Everywhere the testimony is that nine-tenths of all cases of vagrancy, disorderly conduct, and breaches of the peace are the direct effects of intoxicating drinks; hence 3,042 of the 3,380 cases of these offences were due to drink. These, added to the cases of intoxication, will give a total of 6,726 cases, or 88 per cent., as the direct results of the liquor-traffic.

These startling facts deserve and demand the consideration of every one in the community, and should particularly impress our legislators with the necessity of adopting such measures as will tend to change this sad and terrible state of affairs, if not for the sake of humanity, at least for the financial interests of the country. If it costs so much to support our helpless, poor, and criminal population, the State should take the means to prevent and correct these evils.

The Philadelphia County Prison Report for 1871 says, page 16: "About the usual proportion of commitments for the past year may be placed to the account, either directly or indirectly, of intemperance. There were for *intoxication* 3,684, against 3,983 for 1870, 3,546 for 1869, and 2,025 for 1868; for vagrancy, 1,059, against 1,377 for 1870, 1,248 for 1869, and 1,093 for 1868; for assault and battery, 1,821, against 1,376 for 1870, 1,687 for 1869, and 1,462 for 1868; for disorderly conduct and breach of the peace, 2,321, against 5,393 for 1870, 7,360 for 1869, and 8,132 for 1868; for assault with intent to kill, 153, against 132 for 1870, 146 for 1869, and 121 for 1868. Of the entire number of commitments (13,171), nearly three-fourths, or 9,038, are traceable to intemperance; drunkenness being, with exceptions, a cause of the offences in the foregoing list. The aggregate of these offences is considerably smaller than for the two preceding years, it having been in 1870 12,266 and in 1869 13,987. The falling off is chiefly in commitments for breach of the peace —a form of commitment which has to some extent been abandoned by COMMITTING MAGISTRATES under instructions from the Court of Quarter Sessions. It would be unfair to assume that the offences alluded to are exclusively attributable to intemperance; for crime and vagrancy and prisons are found in countries where drunkenness is comparatively rare. But it cannot be doubted that the unrestrained multiplication of temptations to crime in the unbridled sale of alcoholic drinks in our city is a fearful evil."

Mr. William J. Mullen, the well-known and highly-esteemed prison agent, in his report for 1870 says: "An evidence of the bad effects of this unholy business may be seen in the fact that there have been *thirty-four* murders within the last year in our city alone, each one of which was *traceable to intemperance;* and one hundred and twenty-one assaults to murder proceeding from the same cause. Of over

38,000 *arrests* in our city within the year, *seventy-five per cent.* of this number were *caused by intemperance.* Of the 18,305 persons committed to our prison within the year, more than two-thirds were the consequence of intemperance. Of this number, 2,517 *were for intoxication.* The whole number committed to our prison for the offence of drunkenness for the last twenty years was 184,966 persons.

"The whole amount of blood-money which has been paid to our STATE TREASURER for the year 1869 for license to sell intoxicating liquors in this State was $329,211 77, of which over $200,000 was paid by our city for the privilege of contributing nearly a million and a half of dollars for the support of our criminals and pauper population, who are made such by the *use of intoxicating liquors.* If we add to this a fair proportion of the expenses of our charitable as well as criminal institutions of Philadelphia (a large proportion of which is in consequence of intemperance), we have an expenditure of over $2,500,000." Again Mr. Mullen says: "Ignorance and drunkenness are the real causes of nearly all the misery in the world. The last is immeasurably worse than all others combined; for such is the benumbing, stultifying, and crazing effect of inebriating drinks that they change a man of reason and feeling into a brutalized monster. Hence it is that the 'knife, the dagger, the bludgeon, and the pistol are in such frequent use; and in the domestic circle cruelty to children, wife-beating; and in many families at home horrors of every kind.' This is lamentably too true, as is proved by the cases that consume the time of our CRIMINAL COURTS, and is seen by the condition of society at large. No sooner have our courts disposed of one case of murder or assassination than the liquor-shops furnish others to supply its place."

INTEMPERANCE AND CRIME—PHILADELPHIA.

Judge Allison, in a speech delivered at a public meeting in November, 1872, speaking of the evils of intemperance and the duty of good citizens to join in the efforts made to do away with the evils of rum-selling and rum-drinking, said : " Intemperance is upon our right hand and left ; on the streets, north, south, east, and west, we see the lures to destruction, and see that in this city to-night men are being hurried to the drunkard's grave and the drunkard's doom. Shall we be held guiltless if we do not stretch forth our hands and use the means we possess to save our perishing fellow-men? There is a day coming when this question cannot be evaded, but must be answered before an impartial Judge. The lives of these poor drunkards will then be in some measure chargeable to us. There are few people who see the practical evil as we see it in the criminal courts of this city. There we can trace four-fifths of the crimes that are committed to the influence of rum. There is not one case in twenty where a man is tried for his life in which rum is not the direct or indirect cause of the murder. *Rum* and *blood*—I mean the shedding of blood—go hand in hand.

"Shall we not attempt to remedy this thing? Or shall we close our eyes while the agencies for the sale of rum are multiplied? Rum is already a mighty power in this city, and it requires all the power of temperance men to put the traffic under bonds."

The Citizens' Association of Pennsylvania, in their report for 1868, estimated, as already seen, that the number of inmates in the county jails of Pennsylvania was 8,447; of penitentiaries, 669; or an aggregate in both classes of prisons of 9,116, or one person in prison for every 402 of the population. The average cost of maintaining these prisoners was 44 cents per day for each, or a total of $1,464,029 per year, being a cost of $2 45 a year to each voter in the State.

Two-thirds of this cost of crime is estimated by the Citizens' Association to be caused by intemperance.

The number of arrests by the police of Philadelphia in 1872-3, by the reports of Kennard H. Jones, Chief of Police, to his Honor Mayor Stokley, was as follows:

	1872.	1873.	Decr as?, 1873.
Total arrests	40,007	30,400	9,607
Assaults and battery..................	2,358	2,006	352
Assaults with intent to kill...........	205	139	66
Breaches of peace...................	4,661	4,030	631
Intoxication.........................	15,782	10,077	5,705
Intoxication and disorderly conduct....	9,769	7,897	1,872
Total cases usually caused, directly or indirectly, by liquor..............	32,775	24,149	8,626

Total decrease for intoxication and intoxication and disorderly conduct, 7,577.

For the year 1873 there was a decrease for all offences of 9,607; while for the five classes of offences given there was a decrease of 8,626, leaving only 981 for other offences.

By these reports we find the whole number of offences in 1872 was 40,007, of which there were for assault and battery, 2,358; assault with intent to kill, 205; breaches of the peace, 4,661; intoxication, 15,782; and for intoxication and disorderly conduct, 9,769—a total of 32,775 which are directly or indirectly chargeable to the sale and use of intoxicating drinks; the 15,782 cases of intoxication, and 9,769 for intoxication and disorderly conduct, or 25,551 cases caused directly by the use of drink, being 63.86 per cent., or nearly two-thirds, of all the police cases for 1872.*

Besides these prisoners, there were 59,674 lodgers at the different station-houses during 1872.

* By examining the reports of Mayor W. S. Stokley for the years 1872-3 it will be found that of the 40,007 arrests made by the police in 1872, 13,451 were natives of Ireland; and of the 30,403 arrested in 1873, 13,351 were the same nativity. We also find the arrests for assaults, breaches of peace, intoxication, and disorderly conduct in 1872 were 32,775, but in 1873 only 24,149, or 8,626 less than in the previous year; and that there were in 1873 5,110 less Irish arrested than in 1872, while there was only a decrease of 3,597 Americans, 577 Germans, and 325 English. Thus we find that about five-eighths of the decrease of arrests are those of Irish.

Why this difference? By the report of the Catholic Total Abstinence Union's fourth annual convention, he'd in Chicago, October 7 and 8, 1874, on page 27, we read as follows:

"In October, 1872, the Diocesan Union of Philadelphia was composed of five societies, with an aggregate membership of 1,100. In October, 1873, it had increased to 29 societies, 20 adult and 9 cadet, with an aggregate membership of 8,048. At the present date it numbers 57 societies—34 adult and 23 cadet—with an aggregate membership of 12,285, of which 8,577 are adults and 3,709 cadets. This is an increase of 14 adult and 14 cadet societies during the past year, and an increase in the aggregate membership of 4,238. There are 1,100 members of non-union societies, making a grand total of 13,385 total abstainers in Philadelphia."

Is it not clear that this decrease of arrests of Irishmen is due mainly to the existence of the total-abstinence societies of Philadelphia?—for as they have increased in number and aggregate membership, the arrests chargeable directly to intoxicating drinks have proportionately decreased, which is another proof that intoxicating drinks cause crime, while total abstinence decreases it.

It is very safe to say that not less than 40,000 of these lodgers were brought to the necessity of seeking shelter in a police-station by the use of drink.

Besides the cases before our police courts, brought there directly through drink, at least three-fourths of the remaining cases were indirectly caused by liquor. Of the 23 murders that were committed in Philadelphia in 1872, 20 of them, at the lowest calculation, sprang directly or indirectly from the same direful cause. The police expenses were $1,246,713 98. Of this sum, two-thirds would not be needed if the drink-traffic did not exist. To this must be added the expense of building the new House of Correction, which has cost the city already over $575,000. The House of Correction would not be needed but for the use of strong drinks ; for, by the reports of the officials of the institution, over 80 per cent. of the inmates were brought there by intemperance.

Coroner Brown of Philadelphia, in his report for the month of September, 1874, gives the following cases of violent deaths : Mary Heron, thrown downstairs ; Mrs. Tozier, shot by her husband ; Elizabeth Carton, beaten to death by her husband ; Simon Schmid, struck in the head with a beer-glass. All resulted directly from drink, except the case of shooting. Thus 3 out of 4 violent deaths in the city of Philadelphia, in the space of one month, were caused directly by drink.

The Grand Jury for the December term, 1874, of the Court of Quarter Sessions of the City of Philadelphia, in the final presentment, said they "had acted upon 471 bills, of which 324 have been returned as true bills, and 147 have been ignored.

"A large proportion of the cases before us were for assault and battery, and in every instance these were the direct results of a free and improper use of intoxicating drinks. Indeed, this liquor-traffic is the fertilizing source of all crime. It is evident that in a community where a considerable proportion of the people are unable from various causes to resist the temptation which beguiles them at every corner, there should be proper safeguards as a defence for the weak ones. In the protection of society from the devastations of this river of fire, it may yet be necessary to hold the liquor-seller to a criminal responsibility for the crimes committed under the influence of liquors sold by him or them.

"Society must be protected, purified, and elevated from present conditions by wise, intelligent, and far-reaching agencies, religious, social, and legislative. It is a noticeable fact that a very considerable number of these crimes were committed on the Sabbath day; so that the historic consequences which in all ages have followed Sabbath desecration are ripening their poison-fruit in our midst. Statistics well kept constantly show that no legislation of city or State, no social or human contrivance, can for a moment arrest the certain punishment which marches like an armed giant in the path of an ever-present divine retribution. The Sabbath of God cannot be desecrated with impunity by either individuals, corporations, or governments.

"A growing evil and fruitful source of crime in our city arises from the thousands of idle, vagrant youth who wander about the city and congregate in dens of infamy. These are the products, for the most part, of broken and disrupted families, shattered and consumed by the liquid fires of rum. This is a dangerous element in our midst, young, vigorous, and, to some extent, equipped. The well-being of our city imperatively demands the instant suppression of the dens where these youths are harbored and the lowest instincts ministered to and trained to crime. It is clear that when, from crime or other causes, the parent ceases to control or to provide for, educate, and properly train the child, then the State or city government becomes of right and duty the parent, and is bound to enter fully into all the responsibilities and relationship of parent to child. What, then, shall be said of the city parent, rich in palace homes, and overflowing with wealth and prosperity, yet with 15,000 of her youth beggars, thieves, homeless? The only remedy at our hand is COMPULSORY EDUCATION; not a house of correction, but a school. Ignorance is very expensive; crime still more so. Juvenile crime is the most expensive. In a mere dollar sense it would cost much less to the taxpayer to arrest, confine, and educate into societary salvation these children of the street and den than it now does under the present conditions. These wretched outcasts are the city's children."

By the report of the Board of Public Charities of Pennsylvania for 1872 we learn that of the 240 inmates in the Eastern Penitentiary, the habits of 75 were sober, 66 moderate drinkers, and 99 intemperate. Of the 213 inmates of the Western Penitentiary, 63 are reported as sober in their habits, 60 as moderate drinkers, and 90 intemperate. Nor is this merely an accidental proportion of one year; for by the report of 1870 we find that for the nine years from 1860 to 1869 there were sent to the Western Penitentiary 1,500 convicts, whose habits are reported as follows:

Abstainers from liquors, ... 589
Moderate drinkers, ... 274
Intemperate, ... 637
                                          ———
Total, ... 1,500

By the report of 1871 we find that in 1870 144 convicts were sent to that institution, whose habits were given as—

Abstainers from liquors, ... 39
Moderate drinkers, ... 55
Occasionally intemperate, ... 17
Intemperate, ... 33
                                          ———
Total, ... 144

In the Eastern Penitentiary, the same year, there were 315 convicts; their habits are given as—

Abstainers from liquors, ... 24
Moderate drinkers, ... 210
Occasionally intemperate, ... 16
Intemperate, ... 65
                                          ———
Total, ... 315

By the report for 1872 of the Board of State Charities of Pennsylvania, the numbers of convicts sent to Eastern and Western Penitentiaries were as follows:

Habits.	Western. Penitentia.y.	Eastern Penitentiary.	Total.	Per cent. in both.
Sober, ...	63	75	138	30.46
Moderate, . . .	60	66	126	27.82
Intemperate, . . .	90	99	189	41.72
Total, . . .	213	240	453	

By the report of 1871 of the Western House of Refuge, of the 224 inmates, the parents of 76 were intemperate. The report of the State Charities, on page 91, says:

"We have spoken of intemperance as a fruitful source of pauperism and crime, and it is doubtless the proximate cause of nine-tenths of the idleness, brutality, and vice which affect society." The report might with equal truth have added that it was the cause of four-fifths of all crime. The figures given above of the convicts to the penitentiaries you may say do not show so large a proportion of intemperate. But it must be remembered that we cannot by the convicts sent to the penitentiaries arrive at the amount of crime directly or indirectly the result of intoxicating drinks. These criminals are for a different class of crimes, and are an entirely different class of persons from those sent to our county prisons. The class of criminals, as burglars, gamblers, counterfeiters, etc., who require a steady hand and a clear brain to be able to pursue their avocations successfully, must abstain from drink; and these are the abstainers and moderate drinkers reported. The convicts sent to the penitentiaries are not one-tenth of the criminals committed, and it is safe to assert that not less than three-fourths of all the crime committed

in the State of Pennsylvania, and in every other State
in the Union, is directly caused by drink. Again, the
greater amount of crime caused by intemperance is
never brought into our courts, but is settled before it
ever reaches them. Hence it is impossible to fully
estimate the vice and crime caused in any community
by strong drinks, and any statistics we may be able to
collect on the subject will fall short of the truth ; for it
is utterly impossible to estimate the extent of the vice
and crime directly caused by the use of alcoholic
drinks, and much less what is indirectly the result of
the same cause. The city of New York is perhaps
not behind any city in the Union for its number of
liquor-shops and the results following from them, of the
extent of which we may form a slight idea when we
reflect that there were, in 1867, 5,203 places where
liquor was publicly sold, each of which receives the
daily average of 134 visits. These visits are not imagi-
nary; for Superintendent Kennedy placed police-officers
over 223 licensed liquor establishments to observe how
many entered into those places, when it was found that
the average visits to each were, as already said, 134
daily, or 218,224,226 visits annually. The number of
arrests by the police for the year ending October,
1868, was 98,861, of which 50,844 were for intoxication
and disorderly conduct. In addition to the licensed
liquor-shops, there were 647 houses of ill-fame and
1,678 billiard-saloons. Mr. Oliver Dyer, in a lecture
on " The Wickedness of New York," said the liquor-
shops of New York would line both sides of a street
running from the Battery out eight miles into West-

chester County. The Commissioners of the Metropolitan Police reported for 1867 that there were 80,532 arrests, 21,589 of whom were women and 58,948 men. Of the women arrested, 1,056 were for assault and battery, 62 for felonious assaults, 6 for robbery, 6 for murder, 7,529 for disorderly conduct, 4,075 for intoxication, 3,294 for intoxication and disorderly conduct, 1,199 for petty larceny, and 491 for grand larceny. Does not this plainly show the demoralizing influence of strong drinks? Of the men arrested, 17,604 were for intoxication and 13,233 for disorderly conduct. These men were not all uneducated and of the criminal classes; for amongst them were 30 editors and 8 clergymen. Taxpayers, are you willing to pay your hard-earned money to support a system that produces so much crime? Christian men and women, can *you* longer give countenance and support to so great a sin-engendering cause as the traffic in strong drinks? Can you stand idly by and do nothing to free our country from this blighting, withering curse? The arrests in the city of New York in 1871 numbered 75,692; 34,696 were for intoxication and disorderly conduct, the direct consequence of the millions of dollars expended for intoxicating drinks. Besides these arrests, there were 141,780 persons who lodged at the different lodging-houses. It can hardly be supposed that this vast multitude would need to seek shelter in such places if the millions spent for liquor had been kept in the pockets of those who spent them, or if they had been expended for useful and necessary products of our industries; neither would there have been one-tenth of the arrests made. There is certainly a defect in the social system, something radically wrong in our government, that such fruits should be produced. Thirty-four thousand drunken persons arrested in one year for that vice in one city alone, with tens of thousands wandering about the streets with no place to rest their weary heads, which must keep an army of upward of three thousand police to look after these poor victims of the rum-traffic, for which are paid nearly three and a half millions of dollars a year! And still the cry is for more houses to shelter the homeless. Two million dollars are spent annually by the State Board of Charities and Correction. Of the 24,166 persons relieved out of the streets of New York, sixteen thousand were children. The average population of the New York hospitals, asylums, nurseries, prisons, reformatories, etc., is 8,840. Nor do matters grow better in this respect, as is evidenced by the report of Commissioner Stern, which was adopted January, 1874, by the Board of Charities and Correction of New York. By this report we are informed that the number of persons committed to the workhouse on Blackwell's Island more than five times for intoxication, from January 1, 1870, to January 1, 1874, was as follows :

Males Committed.				Males Committed.		
103	. .	6 times before.	1	. .	30 times before.
28	. .	7 " "	2	. .	50 " '
162	. .	8 " "	1	. .	40 " "
5	. .	9 " "	1	. .	70 " "
1?1	. .	10 " "	1	. .	75 " "
16	. .	12 " "	1	. .	80 " "
21	. .	15 " "	1	. .	100 " "
27	. .	20 " "			
4	. .	25 " "	550 total.		

Females Committed.				Females Committed.		
3,702	. . .	6	times.	10	. . .	25 times.
602	. . .	7	"	5	. . .	26 "
1,437	. . .	8	"	1	. . .	28 "
172	. . .	9	"	1	. . .	29 "
1,157	. . .	10	"	36	. . .	30 "
31	. . .	11	"	91	. . .	40 "
749	. . .	12	"	1	. . .	41 "
13	. . .	13	"	1	. . .	48 "
46	. . .	14	"	1	. . .	49 "
37	. . .	15	"	14	. . .	50 "
28	. . .	16	"	1	. . .	58 "
7	. . .	17	"	19	. . .	60 "
33	. . .	18	"	1	. . .	70 "
5	. . .	19	"	1	. . .	80 "
762	. . .	20	"	1	. . .	86 "
1	. . .	21	"	29	. .	100 "
7	. . .	22	"			
1	. . .	23	"	9,006 total.		
3	. . .	24	"			

Meyer Stern, the commissioner, says :

"That account speaks volumes for itself. The tale it tells of male drunkards being recommitted to prison from one hundred times down to six times, of whom one hundred and eighty-one offenders were recommitted ten times, is dreadful to contemplate. But this tale of horror is put entirely in the shade —it is lost sight of—if placed side by side with the statistics of female arrests. While 560 male persons were committed for intoxication during the past three years, there were arrested 9,006 females —sixteen times as many. Of the former, one was rearrested one hundred times for the same offence ; of the female drunkards, twenty-nine had to be rearrested one hundred times ; and this fearful proportion is observed all through. Is not this sufficient evidence of a deplorable defect in the present law, and which we must by all means try to remedy?" Fellow-citizens, it is by your will these things exist. You are the sovereigns; the power is in your hands to remove or still keep this terrible drink-shop system, that may make your sons, your daughters, and your wives, ay, yourselves, equal to the worst of the poor victims of the poisonous cup that were brought down to occupy the cells of Blackwell's Island Workhouse for the hundredth time. Oh! think of New York City alone, with its 5,203 licensed liquor-dens, and perhaps as many more unlicensed; also of its 40,000 destitute, outcast, homeless children; of its 647 houses of ill-fame; its 6,929 cases of assault and battery by men and women; of the 98,861 arrests, nine-tenths of which are the result of drink.

This crime and degradation is not confined to New York alone or Philadelphia; all over our fair land intoxicating drinks are breathing their terrible upas breath, blasting all that is fair or lovely. Nine-tenths of all the crime, the vice, and degradation of our country are chargeable to strong drink. Bronning, the Boston wife-murderer, confessed that he beat his wife to death because she would not give him her hard earnings to spend for drink. Mr. Edmund, warden of New York City prison, said three-fourths of all offences are directly or indirectly caused by intoxicating drinks. Oscar Tyler, sheriff of Albany, said eight-tenths of persons committed to Albany county jail were in consequence of the use of liquors. Seth Clarks, jailer of Buffalo, said nine tenths of the crime in that county had its origin in intemperance. J. C. Cole and S. H. H. Parsons, police justices of Albany, said that three-fourths of all offences are the result of the use of liquors. So we may pass from county to county, from State to State, and the answer from all will be that from four-fifths to nine-tenths of all criminal and other offences are caused by strong drink.

This relation of the use of intoxicating drinks to the production of crime is not accidental, but the direct and essential result of their nature and inevitable tendency. The mass of crime produced by the use of drink was not committed by persons in a positive state of drunkenness, but by far the greater part when the person was just sufficiently under its influence to arouse the lower passions and propensities to the degree when men are easily tempted to do evil and readily provoked to acts of violence, who, but for the excitement of the liquor, would have been able to resist the impulse to do wrong.

It is in the blunting of the mental and moral faculties of man, and in exciting the passions, that the triumphs of drink consist. Burke, the notorious Irish murderer, said he never felt remorse of conscience but once; when about to kill an infant, it smiled in his face. That smile of innocence touched his stony heart. He could not perpetrate the cruel act. But he drank a glass of brandy. That one glass stifled his conscience and blunted all feelings of pity; he then committed the cruel act without pity, without remorse. Bishop and his partner in crime, before they undertook to murder the Italian boy, prepared themselves by imbibing plentifully of gin. Few indeed are the criminals, in this or any other country, who have not had to charge the use of alcoholic drinks, directly or indirectly, with being the cause of their ruin.

CHAPTER XII.

THE success of a republican government depends mainly upon the education of the people. Unless the citizens are intelligent, a free government is always in danger. It is the character of the citizens that makes states and unmakes them ; and as character is mainly formed by education, it is of first importance that all should be well educated. No matter how little or how much we have investigated the subject, this truth meets us everywhere.

The founders of this Republic no doubt felt the need of right education ; and had not slavery existed, they would undoubtedly have inaugurated a general system of education under the control of the General Government. But as this, under the circumstances, could not be done, it was left to the regulation of the several States. Hence the effect of this policy is now very readily seen in the general characters of the natives of the different States in the Union. The States first to adopt the free-school system are among the most prosperous, and their citizens the most wealthy, sober, intelligent, moral, religious, and happy. In most States now provisions are made for the general education of the people and free

schools for the children; yet, with all the means that
have been adopted, there were, in 1870, 4,528,084 persons, ten years old and upwards, who could not read,
and 5,658,144* who could not write; and what is the
most startling in these figures is that 4,880,271 of
these persons who cannot write, over ten years of
age, are *natives* of the *United States.* The foreign-born
who cannot write are 777,873. True, our educational
system is not complete or perfect, and there is room
for much improvement; yet, all circumstances considered, it is equal to that of almost any other
country. Notwithstanding all our educational advantages, and that our schools are multiplying
yearly, there is an almost equal demand for jails
and penitentiaries; and prison statistics inform us
that the inmates of these institutions are yearly increasing. What is the cause of this? Go where you
will, in almost every State there are two entirely
opposite systems of education; and though each
antagonizes the other, yet both are established by
law, and both are fostered and encouraged by the
customs of society, and more or less by the people
of every rank and station throughout the nation.

Both these systems of education cost vast sums of
money to support them. The first of these is the
free or public school, which is the embodiment of the
enlightened ideas of the learned and good of all ages,
and whose nature, tendency, and results are to produce citizens of superior character, to promote the

---

* Compend. Ninth Census Report, page 456.

welfare of all, to build up and strengthen the power and influence of the State.

The other system of education is embodied in the dram-shops of our country, that are now so flourishing under the protection of our Christianized and civilized governments, and licensed by them to educate and make worthless citizens, spread over the States a deluge of corruption and death, destroying the influence of the former system of education or changing much of the good produced by it to evil.

There are in the United States 141,629 schools,* with 221,042 teachers and 7,209,938 pupils, costing $95,402,726. Of these schools, 125,059 are public schools, with 183,193 teachers, 6,228,060 pupils, costing $64,030,673.

Of the population of the United States,† 12,955,443 are between the ages of 5 and 18 years—the school age; but we find that only 6,228,060, or little more than half of those of school age, attend the public school, and only 7,209,938 attend all the schools, colleges, etc., in the country, leaving 4,845,505 of school age who do not attend school at all. Why are they not at school? Some of them may be engaged in various occupations. True, 739,640 between 10 and 15 years are engaged in labor of some kind. But it is safe to say that more than three millions who ought to be at school are not engaged in any occupation of benefit to themselves, their parents, or the state. What are the parents of these children? To whom do those dirty, ragged, and forlorn-looking children belong who are running about the streets, alleys, and by-places of our cities? Ninety-nine hundredths of them are children of the intemperate, who have no care either for their bodies or their souls; their only desire being to obtain drink to stupefy their senses to forgetfulness. But are these children uninstructed? Alas! for them and the welfare of society, no; they are early educated in the schools of crime and nurseries of depravity—the streets. They are taught to live by begging or theft, and before they are fairly in their teens are adepts in crime, steeped in depravity and sin, and soon become graduates in those high-schools and colleges of sin and debauchery—the liquor-shops and low dance-houses.

Our schools and colleges will be inoperative and fail to elevate our people so long as these schools of vice and nurseries of crime—the drinking-shops—are allowed on every hand. For the drink-shops not only close the doors of our schools against the children, but they destroy much of the good produced by the schools by deadening the intellects of our citizens and rendering education and knowledge that has been acquired useless. We may build our school-houses on the most improved style of architecture, and place in them the best teachers that the highest salaries can command; and, when all is done, we shall fail to educate and produce citizens that are industrious, intelligent, honest, and virtuous, so long as the other system of education is allowed to exist, and the law sanctions and allows the schools of immorality and vice to stand side by side with our free schools and colleges. What egregious folly for grave legislators to enact one law to educate our children in science, knowledge, and virtue, and another set to undermine and destroy the benefits of the first! Why should we take such pains and go to so much expense to do good, and then at much more expense to undo the good done? This table of schools, etc., in the United States is compiled from the Census Returns of 1870* and Internal Revenue Report, and includes statistics of all classes of schools, 1870.

* Census Report, 1870, page 450.

TABLE XXII.

	No. Schools	No. Teachers.	No. Pupils.	Total Cost or Expenses.	No. Retail Liquor-sellers.	Cost of Liquors in States and Territories.
				Dol'ars.		Dol'ars.
United States, .	141,620	221,042	7,209,968	95,402,726	143,115	715,575,000
Alabama, . .	2,969	3,364	75,866	976,351	1,976	9,880,000
Arizona, . .	1	7	132	6,000	119	595,000
Arkansas, . .	1,978	2,297	81,526	681,962	2,000	10,000,000
California, . .	1,548	2,444	85,507	2,946,308	5,845	29,225,000
Colorado, . .	142	188	5,033	87,915	371	2,225,000
Connecticut, .	1,917	2,926	93,621	1,856,279	3,352	16,760,000
Dakota, . .	35	52	1,255	9,284	83	410,000
Delaware, . .	375	510	19,575	212,712	368	1,840,000
District of Columbia,	313	573	19,503	811,242	1,087	5,435,000
Florida, . .	377	483	14,670	154,569	580	2,900,000
Georgia, . .	1,880	2,432	66,150	1,253,299	2,707	13,835,000
Idaho, . . .	25	33	1,208	19,988	244	1,220,000
Illinois, . .	11,835	24,056	767,775	9,970,009	8,565	42,825,000
Indiana, . .	9,073	11,652	464,477	2,499,511	4,444	22,220,000
Iowa, . . .	7,496	9,319	217,654	3,570,093	3,073	15,365,000
Kansas, . .	1,689	1,955	53,882	787,226	1,117	5,585,000
Kentucky, . .	5,149	6,346	245.139	2,533,429	4,761	23,805,000
Louisiana, . .	592	1,902	60,171	1,190,684	4,414	22,070,000
Maine, . .	4,723	6,986	162,636	1,106,203	843	4,215,000
Maryland, .	1,779	3,287	107,384	1,993,215	4,285	21,425,000
Massachusetts, .	5,726	7,561	269,337	4,817,939	5,039	25,195,000
Michigan, . .	5,595	9,559	266,627	2,550,018	5,020	25,100,000
Minnesota, . .	2,479	2,836	107,266	1,011,769	1,931	9,655,000
Mississippi, . .	1,564	1,728	48,451	780,331	1,807	9,035,000
Missouri, . .	6,750	9,028	370,337	4,340,805	5,888	29,440,000
Montana, . .	54	65	1,745	41,170	440	2,445,000
Nebraska, . .	796	840	17,614	207,560	635	3,175,000
Nevada, . .	53	84	2,373	110,493	658	3,290,000
New Hampshire, .	2,542	3,355	64,677	574,898	1,101	5,805,000
New Jersey, .	1,893	3,899	120,800	2,982,250	5,649	28,245,000
New Mexico, .	44	72	1,798	29,886	418	2,090,000
New York,	13,020	23,918	863,022	15,936,783	21,318	105,290,000
North Carolina, .	2,161	2,692	64,958	635,892	1,315	6,575,000
Ohio, . . .	11,952	23,589	790,795	10,244,644	11,709	58,805,000
Oregon, . .	637	826	32,593	248,022	738	3,690,000
Pennsylvania, .	14,872	19,522	811,863	9,628,119	13,015	65,075,000
Rhode Island, .	56	951	32,596	565,012	727	3,035,000
South Carolina, .	750	1,103	38,249	577,953	1,565	7,825,000
Tennessee, . .	2,794	3,587	125,831	1,650,692	2,684	13,420,000
Texas, . .	543	706	23,076	414,880	2,168	14,840,000
Utah, . . .	267	408	21,067	150,447	128	640,000
Vermont, . .	3,084	5,160	62,913	707,202	540	2,700,000
Virginia, . .	2,024	2,697	60,019	1,155,585	3,311	16,570,000
Washington, .	170	197	5,409	48,302	224	1,120,000
West Virginia, .	2,445	2,838	104,949	698,061	543	2,715,000
Wisconsin, . .	4,943	7,955	344,014	2,600,310	3,864	19,320,000
Wyoming, . .	9	15	305	8,376	236	1,180,000

To illustrate the effects of these two antagonistic systems of education we will examine a few figures to show their operation in Pennsylvania.

By report of the State Superintendent of Common Schools for 1873 there were in the State, in 1872, 14,415 schools, with 699,802 pupils and an average attendance of 464,127, and 7,674 male teachers and 9,110 female teachers---a total of 16,784 teachers. Total expenditures for common, school purposes, $6,620,498 13. These are exclusive of Philadelphia.

For the city and county of Philadelphia for 1872 there were 1,630 schools; the whole number of pupils registered in 1872, 139,924 ; the whole number belonging to the schools at the beginning of the year, 80,364 ; the number of pupils at the close of the year, 84,387 ; average attendance, 72,025. The whole number of male teachers, 78 ; whole number female teachers, 1,552 ; the total expenses for school purposes, $1,576,199 74. The total educational institutions of the State in 1872 were 16,090 ; teachers, etc., 18,783 ; pupils and students, average attendance, 542,076 ; the cost for educational expenses, $8,399,724.

Let us briefly examine the picture of the other educational system. There were licensed in 1872 in Pennsylvania 15,745 of those schools of drunkenness and debauchery, the retail liquor-shops—and 861 wholesale liquor establishments. Allowing that three persons are engaged in each wholesale place, and two in each retail shop, there will be employed in selling intoxicating drinks not less than 31,490 by the sanction and protection of the State. There are, at the lowest calculation, one-half as many unlicensed liquor-shops as there are licensed, or about 7,000, one-half of which, or more, are in Philadelphia. If two persons are employed in each, it will make 14,000 persons more engaged in selling unlawful liquor, or a total of 45,490 liquor-venders. If each liquor-shop has four drunkards and 30 tipplers, we have in the State 94,424 drunkards and 708,180 tipplers. The total in the State is not less than 802,604 males who are drunkards, tipplers, and sots, besides not less than one-fourth as many females who are tipplers and drunkards, and many of them worse than the worst of the males. The direct cost in the State is not less than eighty million dollars ($80,000,000).

RECAPITULATION OF THE TWO EDUCATIONAL SYSTEMS.

Education in Knowledge and Virtue.		Education in Immorality and Vice.	
Schools, colleges, etc., in Pennsylvania,	16,090	Drinking-places in Pennsylvania,	23,606
Professors and teachers,	18,783	Employed in liquor-shops,	45,490
Pupils and students, etc., in regular attendance,	542,076	Tipplers and drunkards,	802,604
Cost for educational purposes in Pennsylvania,	$8,399,723	The direct cost of liquors in Pennsylvania,	$30,000,000

Thus Pennsylvania has only about two-thirds as many schools, academies, colleges, etc., as there are liquor-shops; and more than twice as many persons are employed in dealing out intoxicating drinks in those schools of vice and immorality as are engaged in schools and other educational institutions; and nearly twice the number .of tipplers and drunkards are attending these schools of drunkenness and immorality as attended all the schools, academies, and colleges in the State; and there was spent for those liquid poisons sold in our licensed drunkard-making manufactories alone ten times more than the cost for true educational purposes.

After carefully examining these figures, can any one wonder that, with all our educational facilities, we have hitherto failed to have a community composed of sober, honest, intelligent, and industrious citizens; that the more our educational improvements have augmented, the more have crime and criminals increased?

Though within the last twenty years our teachers have increased from 25 to 30 per cent., and pupils attending schools more than 50 per cent., yet crime has increased about 60 per cent.

Of 626[*] convicts in the Eastern Penitentiary, Pennsylvania, 390, or 62.30 per cent., had attended public schools; 160, or 25.40 per cent., had attended private schools; and 77, or 12.30 per cent., never went to school.

There were admitted in 1868 into the Houses of Refuge, in Pennsylvania, 536 children, whose ages averaged 14½ years, of which 57 did not know the alphabet, 92 knew only the alphabet, 262 could read

---

[*] From a table prepared by John S. Holloway, Esq., Warden of the Eastern Penitentiary of Pennsylvania.

poorly, 21 read well, 246 could not write, 177 wrote poorly, 94 tolerably, and 19 well.

The county superintendents of common schools of 41 counties visited the almshouses and jails of those counties, and found in the almshouses 2,809 persons over ten years of age, of whom 1,181 could not read, 1,189 could read a little, 412 well, and 70 were good scholars. In the jails there were 1,601 inmates, of whom 434 could not read, 540 could read a little, 504 well, and 123 were good scholars. Of the 291 convicts in the Eastern Penitentiary, 62 were illiterate, 24 could only read, 203 could read and write, and 2 well educated; and of the 5,975 convicts received in this prison, 1,210 were illiterate, 1,019 could read only, 3,714 could read and write, and 32 were well instructed.

Is there not a distinct relation existing between ignorance and crime? Tens of thousands of the children in the State of Pennsylvania do not attend school, though ample provisions are made by the State for the education of every child or person between 5 and 21 years. In the city of Philadelphia, out of 150,000 children between the ages of 6 and 18 years, 20,534 attended neither public nor private schools. Of these 20,534 nearly 11,000 were between the ages of 6 and 12 years; showing that it was not really necessary that they should be kept from school in order to earn their support. It may be safe to estimate that there are from seventy-five to eighty thousand children in the State of Pennsylvania who do not attend school.

If these children are left uneducated, the majority
will find their way either to our jails or poorhouses.
Why are they not at school? Nine-tenths, no doubt,
are children of intemperate parents. By the unani-
mous judgment of the officials of juvenile reformatories, 95 per cent. of the inmates of those institutions came from idle, ignorant, vicious, and drunken
homes. Almost all the children are truant from
school at the time of their committal, have been
habitually idlers on the street, or the children of
besotted and ignorant parents.

IN THE HOUSE OF REFUGE, PHILADELPHIA,

The average number of inmates in the year 1870
was 556. The average yearly cost, including all expenses, except those of a permanent character, and
including earnings, was $125 43 each, and, deducting
earnings, $79 75, or a total of $44,340 for the year
1870.

The admissions were 200 white boys, 38 white girls,
53 colored boys, and 21 colored girls.

From the report we find that 5 had used intoxicating drinks, and nearly all tobacco ; 60 had attended
theatres, 90 were truant-players, 7 had been homeless, and nearly all professional idlers ; 217 had attended public schools, but nearly half were confirmed
truants. In many cases the influence of profligate
and intemperate parents corrupted them and gave,
as it were, a sanction to their own vices.

The average number of inmates in the Western
House of Refuge of Pennsylvania in 1870 was 224,

and the per capita cost, including all expenses, excepting those for permanent improvement, was $201.

The number of admissions was 148, viz., 106 white boys and 30 girls ; and 9 colored boys and 3 girls. The average age on admission of all the inmates was 13 years and 3 months. The report * says : "We thus find that in the case of the parents of those admitted, as far as could be ascertained, 76 parents were intemperate, 27 habitually quarrelsome, 15 had been in prison, 17 were paupers, 7 had been separated, 4 had been insane, and all of the parents of these children but 47 could read and write.

"With respect to the early habits, early training and associations of their childhood, it is recorded that 37 had used intoxicating drinks, 74 had used tobacco, 87 visited theatres, 76 were truant-players, 91 had been idlers, 98 used profane language, 16 had no homes, 13 had been previously arrested, 23 had relatives in prison, 87 had been reared in the family, and 41 among strangers. The causes assigned for these vices are idle habits and bad companions."

* From the Report of Board of State Charities of Pennsylvania, 1871.

TABLE XXIII.

Statistics of Juvenile Reformatories in the United States for 1869, compiled from Tables prepared by B. K. Pierce, D.D., to accompany his paper entitled "A View of Preventive and Reformato y Institutions in the United States."

State.	Title.	No. of Officers and Employees.	Average No. of Inmates in 1868-9.	Percentage of				Cost per Capita for year 1869.	Total Expenditure for 1869.
				Parents Intemperate.	No. who used Liquor themselves.	Foreign Born.	Foreign Parentage.		
								Dollars	Dollars.
California	Industr'al School	18	192	6.10	....	26.00	....	188 00	28,195 00
Connecticu't	State Reform School	17	255	50.	....	.02	50.	150 00	53,115 30
Illinois	Chicago Reform School	..	221	47.62	....	....	....	114 00	25,150 00
Indiana	House of Refuge	25	....	....	....	....	14.	150 00	53,016 27
Kentucky	House of Refuge	13	176	19.	....	2.	41.	80 00	24,055 74
Louisiana	House of Refuge								
Ma ne	State Reform School	17	183	20.	....	5.	40.	110 63	26,000 00
Maryland	House of Refuge	16	340½	35.50	29.	6 02	57.	112 83	39,476 03
Massa'setts	State Reform School	40	307	33.	4.	.12	75.	171 60	52,800 00
"	Nauti'l Ref. Sch. (2 ships)	27	264	....	....	7	68.	192 40	51,500 00
"	State Ind. Sch. (for girls)	19	140	50.	....	.016	50.	170 56	23,891 89
"	House of Reformat'on	10	273	26.00	5.	....	....	....	....
Michigan	State Reform School	25	273	16.	....	....	....		41 000 00
Missouri	House of Refuge	24	183	16.	6.	8.	60.	215 27	39,466 77
New Ham'e	State Reform School	26	101	84.75	10.34	....	....	155 00	15,701 00
New Jer ey	State Reform School	10	67	8.	8.	....	....	153 80	10,989 46
New York	Catholic Protectory	26	584	....	....	....	50.	113 00	63,675 00
"	House of Refuge	54	848	....	....	....	....	128 78	109,204 10
"	Juvenile Asylum	53	623	29 15	....	....	....	131 00	82,855 00
"	West'n House of Refuge	27	517	31.30	....	....	....	114 00	80,063 00
Ohio	House of Refuge	22	193	....	20.	....	....		41,743 27
"	State Reform Sch ol	30	330	....	31.	....	....	133 00	43 805 00
"	St. Re. & In. Sc. for girls	8	12	1.	....	....	....		
Penn'vania	House of Ref., white dept	16	501	....	....	....	....	118 00	59,131 00
"	House of Ref., col'd dept	10	118	....	....	....	....	150 42	19 884 99
"	West'n House of Refuge	23	219	....	....	....	....	251 00	56,940 00
Rhode Isl'd	Providence Reform Sch	19	221	23.	....	....	....	196 00	43 380 27
Vermont	State R form School	7	87	....	....	50.	....	194 00	16,877 00
Wisconsin	State Reform Scho l	..	163	27.77	....	....	....	157 00	25 036 00
Totals for 18 States of the United States and 29 institutions in the same.		584	7,407 Total average of Inmates	25 57 per cent. in 18 institutions.	14 per cent. in 8 institutions.	1 per cent. in 10 institutions.	51 per cent. in 10 institutions.	1 66½ per capita average.	1,114,911 65 Total cost for one year.

In view of these facts we can but say that ignorance breeds crime. And, further, it may be said that the classes most widely debauched by drink are those the least taught in letters and in skilled labor, who are by their drinking habits reduced to the deepest wretchedness of poverty, want, degradation, and helplessness. What are we to expect but that in this condition they will betake themselves to lives of vice and crime? And thus they will become, as figures everywhere prove them, the disturbers of the peace, public order, and the dangerous classes in every community. When all this is true of the parents, what can we expect of the children? Must they not be the 95 per cent. of our juvenile offenders? Must they not grow up to fill our jails and prisons?

The education of our rum-shops counteracts, that of the public schools. Massachusetts may be called the pioneer State of the free-school system. She may support three or four hundred families by teaching, and spends perhaps half a million dollars for education; while as many thousand are supported by the drink trade, and four or five millions are spent for liquors. Boston sends about 35,000 pupils to its public schools, and more than that number into the hands of the police and the officials of the almshouses. Official reports prove that one-eighth of its population are degraded by the drinking-houses of Boston, so that they demand public charity or correction.

These results are not confined to Pennsylvania

and Massachusetts, or to Philadelphia and Boston. Every State, city, town, or village in the United States is almost in the same condition of which we have any statistics. In proportion to the liquor drank and intemperance produced, so education is neglected and crime and pauperism exist. Friends of public education and the human race, are not the means to remedy the evils of intemperance well worth your careful consideration with a view to immediate action?

CHAPTER XIII.

EXPERIENCE and observation have demonstrated beyond a reasonable doubt that at least two-thirds of the moral and social evils afflicting society are due to the use of alcoholic beverages. They also neutralize the efforts for the amelioration of the condition of mankind.

Though the efforts and the means that have been used for the religious, moral, and intellectual development of our people have been numerous and important, yet all must admit their disappointment at the results attained. After all, they have been as successful, perhaps, as could reasonably be expected, considering the adverse circumstances and influences by which they have been surrounded.

Notwithstanding churches and schools are spread all over our land, that thousands are employed to preach the Gospel and as teachers in colleges, academies, and schools, and hundreds more to visit people at their homes to distribute tracts and Bibles, and that millions of tracts and thousands of Bibles have been spread broadcast over our country, and the Gospel preached, yet ungodliness, vice, and immorality abound, and thousands are living without Christ or hope in the life to come.

The principal, if not the sole, cause of this state of things is the use of strong drinks.

The liquor-traffic throws temptations in the way of the old and young, and propagates ungodliness, crime, and sin. There is nothing known within the whole realm of science that possesses the power to degrade and demoralize human beings like alcohol. Its essential properties and nature are such as to carry its victims beyond and out of the reach of all good influences. In this power it stands alone. It benumbs the senses of its victims, deprives them of reason, and renders them incapable of rational and religious impressions. Alcoholic drinks and religion and piety are incompatibles; their relation to each other is as fire to water or an acid to alkali. To talk to men and women about the sublime truths of Christianity who are under the influence of strong drink is little better than to "cast pearls to swine." The use of strong. drinks tends to destroy every personal, social, and religious virtue. A learned physician said: "The devil first binds with a hair, and then with a chain." The man who occasionally drinks intoxicants is bound with a hair which soon becomes a chain that cannot be easily broken, but binds him to the chariot-wheels of Satan. Thousands of good men, ay, Christian men, have been ensnared by this tempter ; prophets, priests, kings, and world-renowned conquerors, have fallen by the potent power of strong drink.

How many clergymen of every denomination have been stripped of their divine office and Christian characters by this monster, and have gone down to the drunkard's grave! None are safe who tamper with it. As the poet has said :

" We are not worse at once;
  The course of evil begins so slowly,
  And from such slight source, an infant's hand
  Might stop the breach with clay.
  But let the stream grow wider, and philosophy,
  Ay, and religion too, may strive in vain
  To stem the headlong current."

Strong drink has always prevented the progress of truth and religion in proportion to the extent of its use. It has continually robbed the Christian Church of its converts, and shorn it of much of its power for the pulling down of the strongholds of sin and Satan, and the establishing of Christ's kingdom. Almost every one can call to mind one or more who for a time ran well the Christian's race, but were finally overcome by strong drink. Can we wonder that strong drink should impede the progress of the Gospel when even ministers, to escape from its terrible power, have to seek refuge in an inebriate asylum?

A large number of our criminals and paupers, as well as those who are fast becoming such, have been Sabbath-school scholars, and sometimes teachers.

To what are these results chargeable, if not to alcoholic drinks? They are undoubtedly the chief cause. The testimony of those in positions to know is very clear on this point.

The chaplain of the Leeds (England) jail said "that of 232 prisoners 180 had been Sabbath-school scholars and 28 had been teachers." Another chaplain, in April, 1869, said: "I have in my book the biography of 650 persons, their antecedents, and what led them to sin; and I might mention, for the information of those connected with the Sabbath-school movement, that nineteen out of every twenty Protestants in prison have at some period of their lives been Sabbath-school scholars. Some of them had been teachers in the Sabbath-school fifteen and twenty years; yet that did not save them from a prison-cell. Now, what was the cause of their fall and finding their way to these cells? Drink, almost without an exception."

Are we any better off in this respect than they are in England? No. Go visit our jails. Ask the inmates if they ever attended Sabbath-school? The vast majority will answer, Yes. Ask them, What brought you here? The answer will be, Drink! I was drunk when I committed the deed for which they sent me here.

Go interrogate the inmates of our penitentiaries and almshouses. The answers will be, Drink! drink! When we compare the extent, the power, and influence of strong drinks, and the temptations and allurements of the traffic in them, with the instrumentalities of the Christian Church, we cannot really be surprised at the comparative result, nor that the Church has failed to produce

effects commensurate with efforts made—the importance and inducements of the Christian religion.

TABLE XXIV.

*Religious Statistics of the United States, from the Census Returns of
1870.* *

Denominations.	Organizations.	Edifices.	Sittings.	Value of Church Property.
				Dollars.
All Denominations, . .	72,459	63,082	21,665,062	354,483,581
1. Baptists (Regular), . .	14,474	12,857	3,997,116	39,229,221
2. Baptists, . . .	1,355	1,105	363,019	2,378,977
3. Christian, . . .	3,578	2,822	865,602	6,425,137
4. Congregational, . .	2,887	2,715	1,117,212	25,069,698
5. Episcopal (Protestant), .	2,835	2,601	991,051	36,514,549
6. Evangelical Association, .	815	641	193,796	2,301,650
7. Friends, . . .	692	662	224,664	3,939,560
8. Jewish, ...	189	152	73,265	5,155,234
9. Lutheran, . . .	3,032	2,776	977,332	14,917,747
10. Methodists, . . .	25,278	21,337	6,528,209	69,854,121
11. Miscellaneous, . . .	27	77	6,935	135,650
12. Moravian (Unitas Fratrum), .	72	67	25,700	709,100
13. Mormon, . . .	189	171	87,838	656,750
14. New Jerusalem Swedenborgian,	90	61	18,755	869,700
15. Presbyterian (regular), .	6,262	5,683	2,198,900	47,828,732
16. Presbyterian (other), . .	1,562	1,388	499,344	5,436,524
17. Reformed Dutch in America (late Dutch Reformed), . .	471	468	227,228	10,359,255
18. Reformed Church in United States (late German Reformed), .	1,256	1,145	431,700	5,775,215
19. Roman Catholic, . .	4,127	3,806	1,990,514	60,985,566
20. Second Advent, . .	235	140	34,555	306,240
21. Shaker, ...	18	18	8,850	86,900
22. Spiritualists, . . .	95	22	6,970	100,150
23. Unitarian, . . .	331	310	155,471	6,282,675
24. United Brethren in Christ, .	1,445	937	265,025	1,819,810
25. Universalist, . . .	719	602	210,884	5,692,325
26. Unknown (local), . .	26	27	11,925	687,800
27. Unknown (union), . .	409	552	153,202	965,295

*Census Report (Compend.) for 1870, p. 514.

TABLE XXV.

Religious Statistics of the United States, 1872*.

	Number Clergy, etc.	Church Members, or Congregations, or Parishes.	Sunday-schools.	Sunday-school Scholars and Teachers.	Contributions for Benevolent objects and Church purposes.
					Dollars.
All Denominations, . .	83,637	11,459,534	26,856	3,754,292	47,636,495
Roman Catholics, . .	3,907	3,758,000	. .	300,000	. .
Methodists Episcopal, .	21,234	1,367,134	16,912	1,410,806	8,796,000
Methodists Episcopal, South,	7,586	571,241	. .	. .	2,258,150
United Brethren in Christ, .	1,709	120,445	2,510	135,954	641,849
Other Methodists, . .	10,968	773,125	. .	. .	. .
Free-Will Baptists, . .	1,145	66,909	. .	. .	. .
Regular Baptists, . .	12,013	1,489,191	5,287	498,756	10,497,103
Disciples, . .	1,797	487,223	. .	. .	. .
Mennonites, Tunkers, Winebrenarians, . .	950	88,000	. .	. .	. .
Seventh Day, Six Principle, Antimission, and other Baptists,	700	70,000	. .	. .	. .
Presbyterian Church, United General Assembly, .	4,795	455,378	. .	479,817	9,097,700
Presbyterian Church, South,	1,096	87,529	. .	50,355	1,034,390
Reformed Presbyterian, 3 sects,	197	19,000	. .	. .	. .
United Presbyterians, .	560	71,804	601	52,616	800,001
Cumberland Presbyterians, .	1,314	96,335	518	20,968	. .
Reformed (German), .	526	101,894	1,019	51,210	594,250
Reformed (Dutch), .	566	63,488	. .	51,160	1,227,657
Congregationalists, . .	3,194	312,054	. .	368,937	6,650,814
Protestant Episcopal, .	2,808	224,905	. .	253,584	5,544,575
Lutherans, . .	2,157	495,325	. .	. .	. .
United Brethren (Moravians),	86	15,064	. .	6,120	131,000
Unitarians, . . .	396	30,000	. .	. .	100,000
Christian Connection, .	3,000	300,000	. .	. .	. .
Universalists, . .	630	84,000	. .	. .	200,000
Friends (Orthodox), . .	. .	57,405	. .	66,090	. .
Hicksite and Progressive Friends,	. .	40,000	. .	. .	. .
New Jerusalem Church of Swedenborgians, . .	63	5,000	. .	. .	63,000
Jews, . . .	. .	50,000	. .	. .	. .
Mormons, . . .	. .	50,000	. .	. .	. .
Spiritualists, . . .	. .	100,000	. .	. .	. .
Minor sects not included elsewhere, . .	150	9,000	. .	. .	. .
Deistical, Atheistical, Radical, and Liberal Clubs, . .	. .	. .	. .	. .	. .

By Tables XXIV. and XXV.—Religious Statistics of the United States, from the Census returns of 1870 and other sources, of the different denominations in 1872—it will be seen that there were in the United States 72,450 religious organizations, a church membership of 11,452,534, with 63,082 churches, and 83,637 ministers. We also find that there were in 1872, 3,754,292 Sabbath-school scholars and teachers, and that the whole contributions for church and benevolent purposes, in 1872, amounted to $47,636,495. Though we have not been able to obtain the contribution of some of the minor denominations and the Roman Catholics, yet we may safely estimate the total contribution for these purposes in 1872 did not exceed the sum of $50,000,000, or less than one-thirteenth of the money spent for liquors; and while there were 63,082 churches, there were not less than 241,716 licensed and unlicensed retail drinking-places, and 7,276 licensed wholesale liquor establishments, or a total of 248,992 places where intoxicating drinks were sold, or nearly four liquor-shops for every church.

Daniel DeFoe wrote, two hundred years ago :

"Where God erects a house of prayer,
The devil builds a chapel there.
It will be found upon close observation,
That the latter has the larger congregation."

This is as true to-day as when written, for if each of the licensed and unlicensed retail drinking-shops have only one-half of the average daily visits of the liquor-shops of New York, or receive 67 visits a day, the retail drinking-places of the United States will receive daily 16,194,972 visits, and 5,911,164,780 visits annually.

Again, while only 83,637 ministers are laboring to spread the Gospel of Jesus, to make men and women better and happier during the present life and fit them for that which is to come, a half-million persons are engaged in dealing out strong drink to destroy them body and soul. Is it not a terrible reflection that not less than 6,000,000 of our population visit these schools of debauchery—the drinking-shops—or more than there are adult church members. Of these 6,000,000 worshipers at the shrines of Bacchus, 600,000 of them are drunkards, of whom 60,000 will annually fill the drunkard's grave, and, unless we deny the Bible, we must believe they cannot enter the Kingdom of God. But the terrible sufferings and awful deaths of these 60,000 human beings leave not less the number of poor miserable suffering drunkards. For as fast as the grave closes over the sad remains of one poor victim of intemperance, another recruit is drafted into the ranks of this army of confirmed drunkards, who are marching on, if by fours, in one column extending 174 miles. Think of this awful procession of human beings, four deep, and 174 miles long! Christian men and women, fellow-citizens, behold this mournful procession; listen to their tramp, tramp, tramping, as they march to the final destruction of body and eternal death of their souls! See, as they tramp along, every ten minutes one of them falls before your eyes and sinks into the *drunkard's grave!* And thus they will continue to drink, to drink, and one will die every ten minutes from the beginning of January to the last minutes of the last day of December of every year, and the number will still increase as the facilities for drinking, the drink-shops, increase.

Can we wonder that good men despond? Is it strange and surprising that the church should make such slow progress in the reformation and regeneration of mankind? Not at all. The only wonder is, that so much good has been accomplished, and that the church has made the advances it has in the face of these 248,992 temples of Bacchus, with their 505,000 priests, who are continually employed in dealing out death, ruin, sin, and demoralization. But the God of Daniel still lives, and, if we do our duty, he will deliver us from the jaws of these lions.

As good Christian men and women, we pray to God to revive his work and to pour his Holy Spirit into the hearts of men, that the Gospel of peace and goodwill toward men may be spread over all the earth, that his kingdom may come and his will be done on earth as in heaven. Let us not be deceived. God works by means and not by witchcraft. He expects men and women to co-operate with his power and be instruments in his hands to save mankind from their sins. Strong drink has defied and frustrated the labors of the church of Christ to evangelize

the world, by keeping millions of souls from listening
to the message of the Gospel's glad tidings of great
joy, and deliverance from the thraldom of sin and
death, and still be allowed to fill our streets with
drunkards on the Sabbath.

The Christian church can hardly be expected to
accomplish its God-designed mission while the liquor-
traffic is allowed to exist.

The cases are very few in which persons have been
expelled from evangelical churches that strong drink
was not the direct or indirect cause.

This has been the condition of affairs since John
Wesley, while visiting Newcastle, excluded seventeen
persons from the society for drunkenness. The
Rev. Newman Hall informs us that "the churches
of England lose on an average one member annually
through liquor-drinking," and that "30,000 members
are slaughtered yearly through this cause." Rev.
Richard Knill said : "Nearly all the blemishes which
have been found on the character of ministers for the
last fifty years have arisen directly or indirectly from
the use of intoxicating liquors."

Rev. Dr. Guthrie, of Edinburgh, said: "I have
seen no less than ten clergymen, with whom I have
sat down to the Lord's table, deposed through strong
drink."

Rev. Dr. Campbell, of London, said: "There has
been scarcely a case requiring of me church discipline, such as expulsion, which has not arisen
through strong drink."

Rev. Wm. Jay, of Bath, said : "In one month not

less than seventeen dissenting ministers came under my notice who were suspended through intoxicating drinks.''

In the report, made February 25, 1869, by the Committee on Intemperance for the Lower House of Convocation of the Province of Canterbury, we find the following testimony of the clergy on the effects of intemperance on the work of the church : 472.* ''Truly drink is the curse of the working classes of London.'' 474. ''Habits of occasional intemperance keep men away from church for a time.'' 475. ''The apparent result is chiefly neglect of the means of grace and ordinances of religion.'' 476. '' Public-house keepers rarely or ever come to church.'' 479. ''Those who tipple most are most frequently absent from public worship.'' 480. ''As a rule they neglect the ordinances of religion altogether.'' 481. ''Necessarily injurious to religion.'' 482. ''There are families who never attend divine service ; they plead that they have no decent clothes in which to come—the truth being that the money which should purchase clothes is spent at the beer-shops.'' 484. ''The Saturday evening attendance at the public-house must, as far as it extends, act injuriously on the duties belonging to Sunday.'' 485. ''No drunkard attends the ordinances of religion.'' 486. ''The effect is to lessen the frequency of the attendance at church.'' 487. '' Sabbath-breaking, swearing, and drunkenness are the vices that go together. The

* The figures are the number of the answers to interrogatories sent to the clergy, as given in the report.

influence for evil here is very great." 490. "All persons who frequent ale-houses are irregular in their attendance at a place of worship." 492. "The Gospel fails to meet the case." 496. "One public-house only. Population, 280. Since the opening of the public-house, the attendance at church has been somewhat less." 500. "There is a beer-shop in the next parish, which is a source of annoyance. The license is continued contrary to the wishes of myself and several respectable neighbors. It is very discouraging to the parochial clergy of small agricultural parishes, where their efforts for the spiritual welfare of the people are in a great measure frustrated by the baneful effects of beer-houses in their immediate neighborhood ; often in the most retired and by-places favorable for the resort of the basest characters." 501. "Attendance at church has been greatly increased with the decrease of intemperance among my parishioners."

502. "I speak clerically, and say that intemperance undoes all we can do for the moral improvement of the parish ; and, magisterially, that out of every one hundred cases, ninety at least of the cases brought before the bench are directly or indirectly to be traced to intemperance ; and, perhaps, having been in practice for several years as a medical man, and holding my diploma, I may speak *medically*, that the vice caused to a great extent by intemperance ruinously affects the health of numbers."

503. "Numbers are kept away from public worship from intemperance directly, and still more so

perhaps indirectly, the effect of intemperance being not only to produce poverty, but also to debase and deprave the whole moral nature—in fact to brutalize." 504. "Intemperance keeps numbers from church." 505. "The consequences on morals and religion are clearly marked in both respects." 516. "I believe the beer-houses and publics to be the devil's own hot-houses. . . . . . . Communion has become really extinct among the poor, and vital religion as low as well can be." 517. "The utter annihilation of all moral and religious feeling." 518. "People who indulge in drink seem dead to religion." 519. "No one can well exaggerate the very injurious influences which the public-houses exercise over the religious and moral feeings of my population." 520. "It is always hostile to religion and morality." 522. "It is very prejudicial to religion, more than any other cause, and is the secret source of backsliding among Christian converts." 523. "It need hardly be said that intemperance is the fruitful source of irreligion." 527. "With the increase of beer-shops there has been a decreased attendance on the ordinances of religion." 529. "I never see the habitually intemperate at church. I have often known men change their habits, and then come to church." 530. "Keeps the men from going to church, and renders them indifferent to religion." 531. "A considerable amount of Sabbath desecration consequent on intemperance." 532. "Destroys all regard for religion." 534. "As touching religion the place is demoralized. No one is ashamed of drunkenness, and the violent deaths which not unfrequently occur are no warning. Only a few weeks since a drunken man was roasted to death upon a lime-kiln bank, and the same day his two brothers consoled themselves by a drunken debauch. I have told them at church that drink is the God of S——, and the public-houses their churches." 536. "No room in man's heart for two gods—when they worship drink, there is a corresponding absence from God's worship." 539. "Almost all that is wrong in the parish—wrong and irreligious—is traceable to drunkenness." 544. "Intemperance disqualifies for profitable attendance on religious ordinances." 545. "As soon as a drunkard leaves off drink, even temporarily, he generally begins to attend divine service. If he drinks, he keeps away, except at *club sermons.*" 546. "They become so degraded that they are ashamed to be seen poorly clad in places of worship." 547. "A fearful drawback in morals and religion; it ruins my senior scholars awfully."

This testimony of the clergy of England corroborates what has already been said of the demoralizing and irreligious tendencies of strong drinks. These effects are not confined to England; the same results are produced wherever used, whether on this or the other side of the Atlantic. As early as 1831, the Rev. Mr. Barbour, of New England, set himself to work to ascertain the losses caused to the churches by liquor-drinking. He addressed "circulars" to ministers and clerks of churches in all of the New England States, and of New York,

New Jersey, Pennsylvania, and Ohio. He received replies from 459, whose records show 2,590 cases of discipline where the charge was intemperance alone. From this and other data obtained, he concludes that *seven-eighths* of all cases of church discipline arise directly or indirectly from liquor-drinking. Another gentleman gives the following testimony: "I have travelled in 48 counties, and visited 450 churches in Pennsylvania and in many other States, embracing nearly all denominations. I have made diligent enquiry in regard to drinking by ministers and church members, and these are my conclusions, viz.:

"1. That the churches of this country lose, on an average, one member a year from liquor-drinking.

"2. That liquor-drinking causes the ruin of more ministers than all other causes combined. That a minister rarely falls who is not at least a tippler.

"3. That since 1855, when the slavery agitation broke up our systematic temperance education, drinking customs have increased at least one hundred per cent. in the churches of this country."

These statements are plainly within the truth, as the records of every church in the country will testify. Let any church member or minister examine the records, or call to mind all the cases of church discipline of which he has any knowledge, and he will find that the major portion arose from the use of strong drink. Sufficient testimony has been adduced to leave no doubt in the mind of any person of the injury inflicted upon the Christian Church by strong drinks. This demoralizing traffic must be abolished. The Gospel can never fully spread its soul-saving influence while we have four of these devil's chapels— drink-shops — for every church; and spend one dollar for the spread of the Gospel and Christian charities, and more than thirteen for intoxicating drinks, to spread crime, sin, and debauchery. Strong drink shuts out the Holy Spirit. It stifles the convictions, sears the conscience after the person has been awakened.

Strong drink obstructs the progress of the Gospel. The intemperance of the Christian professors in foreign lands brings reproach upon the holy religion of Jesus. Sir Charles E. Trevelyan, K.C.B., in his testimony to the Committee on Intemperance of the Convocation of the Province of Canterbury, said: 1,211. "The responsibility of the empire has also to be considered. Those only who have lived in heathen countries know what a scandal to our nation and to Christianity will be removed by a change in our military system. The natives of India ask whether the *Gora log* (European soldiers) are the same caste as the *Sahib log* (European gentlemen); and seeing the exhibition our soldiers too often make of themselves in the grog-shops and houses of ill-fame, in the bazaar, they wonder why, if this be the result of a Christian education, the missionaries take such pains to convert the Hindoos and Mohammedans to Christianity. To abstain from intoxicating liquors is a cardinal point of both those religions, and it is a

disgraceful fact that the tendency of our influence has
been to encourage excess in the use of them. We
are not speaking now of money, but of money's
worth; and surely it is worth something, even for the
peace and duration of our Indian empire, so to con-
stitute our military force that it may present the as-
pect of a Christian army to the population of many
races, languages, and religions, whose welfare is dependent upon us." The following is a testimony sent
to the chairman of the Committee by Sir John
Bowring:

"It has been deemed somewhat singular that
neither in the HEBREW nor the CHRISTIAN CODE is the
vice of drunkenness especially censured or forbidden.*
It may be sufficient to reply that if the commands of
the Decalogue or of the all-comprehensive teachings of
Jesus were obeyed, intemperance in any form would be
impossible, and as the greater must include the less,
the highest religious authority is not wanting to discourage the vice of inebriety. Still the lamentable fact
remains that drunkenness is far more common among
nations professedly Christian than among those who
have any other national faith. In the Levant the use
of strong drinks is almost wholly confined to the
CHRISTIAN and the HEBREW races, for though intoxicating liquors are used among the Mohammedans, the
use is *secret*, as public opinion would not tolerate its
public employment. So strong are the prohibitory en-

* Deut. xxi. 20, xxiv. 19; Prov. xxiii. 21; Isa. v. 22, xxviii. 7; Hab. ii. 15;
Matt. xxiv. 49; Luke xii. 45, xxi. 34; Rom. xiii. 13; Gal. v. 21; 1 Cor. v. 11, v. 10;
Eph r. 18.

actments of the Koran that the stricter sects of Mus-
sulmans—such as the Wahabees—will not allow the
use of *coffee*, on account of its exciting qualities. The
value of water as one of the gifts of Allah is constantly put forward in 'The Book,' and the moralists of
Islam all teach that water, which it is permitted to
sweeten with the unfermented juice of fruits or
flowers, is all-sufficient to quenching thirst, and ad-
ministering to unforbidden enjoyment without the
addition of any inebriating element. Water is the
universal drink of Buddhists and Brahmins, and
under these designations we may include nearly half
of the whole race of man. Stimulants of another
character are no doubt largely employed among
Orientals, the hashish of the Arabians, the bang
among the East Indians, the opium among the
Chinese, are very largely consumed ; but, though they
are dangerous to health, and fetch on misery, they do
not generate such seeds of violence, nor lead to sacrifice and suffering, at all comparable in amount or extent to that produced by drinking in the British
dominions.''

Archdeacon Jeffreys, a missionary in the East
Indies, said, more than twenty years ago, ''that for
one really converted Christian, as the fruit of missionary labor, for one person 'born again of the Holy
Spirit, and made a new creature in Jesus Christ'—for
one such person, the drinking practices of the English
had made one thousand drunkards.''

These facts cannot fail to excite painful emotions
in the heart of every Christian. They should arouse

the churches of our country to greater activity, and become united to aid in the overthrow of a legal system that produces so much evil in the Church and out of it. The paramount question with every Christian should be: What is the influence of strong drinks on men's souls? Do they prepare them to receive the Gospel intelligently and reverently, or do they disqualify them to feel and understand its claims and embrace it? The essential characteristics of alcohol are such as to blunt the sensibilities, neutralize the effects of the Gospel, and to disqualify men to accept the offers of mercy and salvation. Can Christians, and especially Christian ministers, be indifferent to the consequences of intoxicating drinks, and the extent to which they defeat their labors.

These evils do not terminate with persons outside of the pale of the Church, whom as Christians we must endeavor to bring into the Church, but the baleful influences of these drinks extend and enter within the sacred limits of the Church, and drag laymen and ministers from the class-rooms and the pulpits into the taverns and beer-saloons, and bring reproach upon religion, and thus do more to retard the progress of the Gospel than the faithful ministers and the whole membership of the Christian Church can do to advance it. Surely, in view of these facts, the Christian and moral people will gravely consider the subject, and unite in the adoption of measures to secure the needed reform.

PUBLICATIONS

OF THE

NATIONAL TEMPERANCE SOCIETY

AND

PUBLICATION HOUSE.

HON. WM. E. DODGE,	T. T. SHEFFIELD,	J. N. STEARNS,
President.	*Treasurer.*	*Cor. Sec. and Pub. Agt.*

THE NATIONAL TEMPERANCE SOCIETY, organized in 1866 for the purpose of supplying a sound and able temperance literature, have already stereotyped and published over four hundred and fifty publications of all sorts and sizes, from the one-page tract up to the bound volume of 500 pages. This list comprises books, tracts, and pamphlets, containing essays, stories, sermons, arguments, statistics, history, etc., upon every phase of the question. Special attention has been given to the department for

SUNDAY-SCHOOL LIBRARIES.

Seventy-seven vols. have already been issued, written by some of the best authors in the land. These have been carefully examined and approved by the Publication Committee of the Society, representing the various religious denominations and temperance organizations of the country, which consists of the following members:

PETER CARTER,	REV. HALSEY MOORE,	REV. J. B. DUNN,
REV. ALFRED TAYLOR,	JAMES BLACK,	REV. R. S. MACARTHUR,
T. A. BROUWER,	REV. A. G. LAWSON,	R. R. SINCLAIR,
J. N. STEARNS,	A. A. ROBBINS,	REV. WM. HOWELL TAYLOR.

These volumes have been cordially commended by leading clergymen of all denominations, and by various National and State bodies all over the land. The following is the list, which can be procured through the regular Sunday-School trade, or by sending direct to the rooms of the Society.

At Lion's Mouth. 12mo, 410 pages. By Miss Mary Dwinell Chellis. **$1 25**

Adopted. 18mo, 236 pp. By Mrs. E. J. Richmond, - - **60**

Andrew Douglass. 18mo, 232 pages, - - - - - **75**

Aunt Dinah's Pledge. 12mo, 318 pages. By Miss Mary Dwinell Chellis, - - - - **1 25**

Alice Grant; or, Faith and Temperance. 12mo, 352 pages. By Mrs. E. J. Richmond, - - - **1 25**

All for Money. 12mo, 340 pages. By Miss Mary Dwinell Chellis. **$1 25**

Barford Mills. 12mo, 246 pages. By Miss M. E. Winslow, **1 00**

Best Fellow in the World, The. 12mo, 352 pages. By Mrs. J. McNair Wright, - - - **1 25**

Broken Rock, The. 18mo, 130 pages. By Kruna, - - - **50**

Brook, and the Tide Turn-ing, The. 12mo, 220 pages, - **1 00**

Come Home, Mother. 18mo,
143 pp. By Nelsie Bruck. Illustrated
with six choice engravings, - **$0 50**

Drinking Fountain Stories,
The. 12mo, 192 pages, - - **1 00**

Dumb Traitor, The. 12mo,
336 pp. By Margaret E. Wilmer, **1 25**

Eva's Engagement Ring.
12mo, 189 pp. By Margaret E. Wilmer,
90

Echo Bank. 18mo, 269 pages.
By Ervie, - - - - **85**

Esther Maxwell's Mistake.
18mo, 236 pp. By Mrs. E. N. Janvier,
1 00

Fanny Percy's Knight-Er-
rant. 12mo, 267 pp. By Mary Graham,
1 00

Fatal Dower, The. 18mo,
220 pp. By Mrs. E. J. Richmond, **60**

Fire Fighters, The. 12mo,
294 pp. By Mrs. J. E. McConaughy,
1 25

Fred's Hard Fight. 12mo,
334 pp. By Miss Marion Howard,
1 25

Frank Spencer's Rule of
Life. 18mo, 180 pp. By John W. Kirton, - - - - **50**

Frank Oldfield; or, Lost
and Found. 12mo, 408 pp., - **1 50**

Gertie's Sacrifice; or,
Glimpses at Two Lives. 18mo, 189 pp.
By Mrs. F. D. Gage, - - - **50**

Glass Cable, The. 12mo, 288
pp. By Margaret E. Wilmer, - **1 25**

Hard Master, The. 18mo,
278 pp. By Mrs. J. E. McConaughy,
85

Harker Family, The. 12mo,
336 pp. By Emily Thompson, - **1 25**

History of a Threepenny
Bit. 18mo, 216 pp., - - - **75**

History of Two Lives, The.
By Mrs. Lucy E. Sandford. 18mo, 132
pp. A tale of actual fact, with an introduction by Rev. S. I. Prime, D D., **50**

Hopedale Tavern, and
What it Wrought. 12mo, 252 pp. By I.
Wm. Van Namee, - - - **$1 00**

Hole in the Bag, and Other
Stories, The. By Mrs. J. P. Ballard.
12mo, - - - - - **1 00**

How Could he Escape?
12mo, 324 pp. By Mrs. J. McNair
Wright, - - - - **1 25**

Humpy Dumpy. 12mo, 316
pp. By Rev. J. J. Dana, - **1 25**

Jewelled Serpent, The.
12mo, 271 pp. By Mrs. E. J. Richmond,
1 00

John Bentley's Mistake.
18mo, 177 pp. By Mrs. M. A. Holt, **50**

Job Tufton's Rest. 12mo,
332 pp., - - - - **1 25**

Jug-or-Not. 12mo, 346 pp.
By Mrs. J. McNair Wright, - **1 25**

Life Cruise of Captain Bess
Adams, The. 12mo, 413 pp. By Mrs. I.
McNair Wright, - - - **1 50**

Little Girl in Black. 12mo,
212 pp. By Margaret E. Wilmer, **90**

McAllisters, The. 18mo, 211
pp. By Mrs. E. J. Richmond, - **50**

Model Landlord, The. 18mo,
202 pp. By Mrs. M. A. Holt, - **60**

More Excellent Way, A,
and Other Stories. By M. E. Winslow.
12mo, 217 pages, - - - **1 00**

Mr. Mackenzie's Answer.
12mo, 352 pp. By Faye Huntington,
1 25

National Temperance Ora-
tor, The. 12mo, 288 pp. By Miss I.
Penney, - - - - **1 00**

Nettie Loring. 12mo, 352 pp.
By Mrs. Geo. S. Downs, - **1 25**

Norman Brill's Life Work.
By Abby Eldridge. 12mo, 218 pp.,
1 00

Nothing to Drink. 12mo, 400
pp. By Mrs. I. McNair Wright, **1 50**

Old Brown Pitcher, The.
12mo, 222 pp. By the author of "Susie's
Six Birthdays," - - - - **$1 00**

Old Times. 12mo, 351 pp. By
Miss M. D. Chellis, - - **1 25**

Out of the Fire. 12mo, 420
pp. By Miss Mary Dwinell Chellis,
1 25

Our Parish. 18mo, 252 pp.
By Mrs. Emily Pearson, - - **75**

Packington Parish, and the
Diver's Daughter. 12mo, 327 pp. By
M. A. Paull, - - - - **1 25**

Paul Brewster & Son. By
Helen A. Chapman. 12mo, 238 pp.,
1 00

Philip Eckert's Struggles
and Triumphs. 18mo, 216 pp. By the
author of "Margaret Clair," - **60**

Pitcher of Cool Water,
The. 18mo, 180 pp. By T. S. Arthur.
50

Rachel Noble's Experi-
ence. 18mo, 325 pp. By Bruce Ed-
wards, - - - - - **90**

Red Bridge, The. 18mo, 321
pp. By Thrace Talman, - - **90**

Roy's Search; or, Lost in
the Cars. 12mo, 364 pp. By Helen C.
Pearson, - - - - **1 25**

Rev. Dr. Willoughby and
his Wine. 12mo, 458 pp. By Mrs.
Mary Spring Walker, - - **1 50**

Seymours, The. 12mo, 231
pp. By Miss L. Bates, - - **1 00**

Silver Castle. By Margaret E.
Wilmer. 12mo, 340 pages, - **1 25**

Temperance Doctor, The.
12mo, 370 pp. By Miss Mary Dwinell
Chellis, - - - - - **1 25**

Temperance Speaker, The.
By J. N. Stearns, - - - - **75**

Temperance Anecdotes.
12mo, 288 pp., - - - - **1 00**

Tom Blinn's Temperance
Society, and Other Stories. 12mo, 316 pp.
1 25

Time Will Tell. 12mo, 307
pp. By Mrs. Wilson, - - **1 00**

Tim's Troubles. 12mo, 350
pp. By M. A. Paull, - - **1 50**

Vow at the Bars. 18mo, 108
pp., - - - - - - **40**

Wealth and Wine. 12mo,
320 pp. By Miss Mary Dwinell Chellis,
1 25

White Rose, The. By Mary
J. Hedges. 12mo, 320 pp., - **1 25**

Work and Reward. 18mo,
183 pp. By Mrs. M. A. Holt, - **50**

Zoa Rodman. 12mo, 262 pp.
By Mrs. E. J. Richmond, - **1 00**

NEW BOOKS.

The Brewer's Fortune.
12mo, 440 pp. By Miss Mary Dwinell
Chellis. **1 50**

Our Coffee - Room. 12mo,
278 pp. By Elizabeth Cotton.
1 00

A Piece of Silver. 18mo,
180 pp. By Josephine Pollard.
50

A Strange Sea Story. 12mo,
427 pp. By Mrs. J. McNair Wright
1 50

Ten Cents. 12mo, 334 pp.
Miss Mary Dwinell Chellis. **1 25**

The Wife's Engagement
RING. 12mo, 278 pp. By T. S.
Arthur. **1 25**

MISCELLANEOUS PUBLICATIONS.

Alcohol : Its Place and Power. By James Miller: and The Use and Abuse of Tobacco. By John Lizars. **$1 00**

Alcohol : Its Nature and Effects. By Charles A. Story, M.D., **90**

Bacchus Dethroned. 12mo, 248 pp. By Frederick Powell, **1 00**

Band of Hope Manual. Per dozen, - - - - - - **60**

Bases of the Temperance Reform, The. 12mo, 224 pp. By Rev. Dawson Burns, - - - **1 00**

Bound Volume of Tracts. No. 1. 500 pp., - - - **1 00**

Bound Volume of Tracts. No. 2. 384 pp., - - - **1 00**

Bound Volume of Sermons, **1 50**

Bible Rule of Temperance, By Rev. Geo. Duffield, D.D., - **60**

Bible Wines; or, The Laws of Fermentation and Wines of the Ancients. 12mo, 139 pp. By Rev. Wm. Patton, D.D. Paper, **30** cts.; cloth, **60**

Bound Volume of Almanac for 1869, '70, '71, '72, '73, '74, '75, '76, **1 00**

Centennial Temperance Memorial Volume. This is a large octavo volume of 1,000 pages, containing the full report of the proceedings of the International Temperance Conference in Philadelphia in June, 1876, and a history of the different temperance organizations in this country and Europe; also valuable essays on almost every phase of the question. Sold by subscription. **5 00**

Catechism on Alcohol. Per dozen, - - - - - - **60**

Communion Wine; or, Bi- ble Temperance. By Rev. Wm. Thayer. Paper, **20** cts.; cloth, - - **50**

Cup of Death, The. A Concert Exercise. 16 pages. By Rev. W. F. Crafts. 6 cts. each; per doz., **$0 60**

Delavan's Consideration of the Temperance Argument and History, **1 50**

Drops of Water. 12mo, 133 pp. By Miss Ella Wheeler, - **75**

Four Pillars of Tempe- rance. By J. W. Kirton, - - **75**

Forty Years' Fight with the Drink Demon. 12mo, 400 pp. By Chas. Jewett, M.D., - - **1 50**

Hints and Helps for Woman's Christian Temperance Work. By Miss Frances E. Willard. 12mo, 72 pp., **25**

Liquor Laws of the United States, - - - - - **25**

Lunarius: A Visitor from the Moon, - - - - - **35**

Medical Use of Alcohol, The. By James Edmunds, M.D. Paper, **25** cts.; cloth, - - - **60**

National Temperance Al- manac, - - - - - **10**

On Alcohol. By Benjamin W. Richardson, M.A., F.R.S., of London, with an introduction by Dr. Willard Parker, of New York. 12mo, 190 pp. Paper covers, **50** cts.; cloth, **1 00**

Our Wasted Resources; or, The Missing Link in the Temperance Reform. By Wm. Hargreaves. 12mo, 220 pp., - - - - - **1 25**

Packet of Assorted Tracts, No. 1. Comprising Nos. 1 to 53 of our list, making 250 pp., - - - **25**

Packet of Assorted Tracts, No. 2. Comprising Nos. 53 to 100, making 250 pp., - - - - - **25**

Packet of Assorted Tracts, No. 3. Comprising Nos. 100 to 150 of our list, making 240 pages, - - **25**

Packet of Temperance Leaflets, No. 1. 128 pp., - - **10**

4

Packet of Temperance Leaflets, No. 2. By T. S. Arthur. 128 pp., - - - - - - - **$0 10**

Packet of Prohibition Documents, - - - - - **25**

Packet of Crusade Documents, - - - - - - **25**

Packet No. 1 of Pictorial Tracts for Children, - - - **25**

Packet No. 2 of Pictorial Tracts for Children, - - - **25**

Prohibition Does Prohibit; or, Prohibition not a Failure. 12mo, 48 pp. By J. N. Stearns, - - **10**

Scripture Testimony against Intoxicating Wine. By Rev. Wm. Ritchie, - - - - - **60**

Temperance Cyclopædia. By Rev. J. B. Wakeley. 12mo, 244 pp., **2 00**

Temperance Lesson Leaves, No. 1, 2, 3, each 8 pp. By Rev. D. C. Babcock. Per 100, - - **$1 00**

Temperance Catechism. Per dozen, - - - - - **60**

Temperance Exercise. By Rev. Edmund Clark, - - - **10**

Text-Book of Temperance. By Dr. F. R. Lees, - - - **1 50**

Two Ways, The. A Concert Exercise. 16 pp. By George Thayer. 6 cts. each; per dozen, - - - **60**

Woman's Temperance Crusade, The. By Rev. W. C. Steele, with an introduction by Dr. Dio Lewis. 12mo, 83 pp., - - - - - - **25**

Zoological Temperance Convention. By Rev. Edward Hitchcock, D.D., - - - - - **75**

PAMPHLETS.

Bound and How; or, Alcohol as a Narcotic. By Charles Jewett, M.D. 12mo, 24 pp., - - - **10**

Buy Your Own Cherries. By John W. Kirton. 12mo, 32 pp., **20**

Example and Effort. By Hon. S. Colfax. 12mo, 24 pp., - **15**

Father Mathew. Address by Hon. Henry Wilson. 12mo, 24 pp., **15**

Illustrated Temperance Alphabet, - - - - - - **25**

John Swig. A Poem. By Edward Carswell. 12mo, 24 pp. Illustrated with eight characteristic engravings, printed on tinted paper, - - **15**

On Alcohol. By Benjamin W. Richardson, M.A., M.D., F.R.S., of London, with an introduction by Dr. Willard Parker, of New York. 12mo. 190 pages. Cloth, $1; paper covers, **50**

Prohibition Does Prohibit; or, Prohibition Not a Failure. By J. N. Stearns. 12mo, 48 pp., - - **10**

Proceedings of National Temperance Conventions held in Saratoga in 1865, Cleveland in 1868, Saratoga in 1873, Chicago in 1875; each, - **25**

Rum Fiend, The, and Other Poems. By William H. Burleigh. 12mo, 46 pp. Illustrated with three wood engravings, designed by Edward Carswell, - - - - - **20**

Scriptural Claims of Total Abstinence. By Rev. Newman Hall. 12mo, 62 pp., - - - - **15**

Suppression of the Liquor Traffic. A Prize Essay, by Rev. H. D. Kitchell, President of the Middlebury College. 12mo, 48 pp., - - **10**

Temperance and Education. 18mo, 34 pp. By Mark Hopkins, D.D., President of Williams College, **10**

5

MUSIC AND SONG BOOKS.

Band of Hope Melodies.
Paper, - - - - - **$0 10**

Bugle Notes for the Tempe-
rance Army. Edited by W. F. Sherwin
and J. N. Stearns. Price, paper, **30**
cts. ; boards, - - - - **35**
Board covers, per doz., - - **4 00**
Paper covers, per doz., - - **3 40**

Campaign   Temperance
Hymns, for Temperance Singers everywhere. 30 hymns, 24 pp. Per 100, **3 00**

Our Songs. 8 pages. Containing 17 hymns suitable for public meetings. Per 100, - - - - **1 00**

Ripples of Song. Price 15 cts.,
paper covers; per 100, **$12.** Board
covers, **20** c:s. ; per 100, - **$18 00**

Temperance Hymns in sheet
form, size 9½x7½ inches, containing
hymns suitable for Public Temperance
Gatherings and Organizations. Price,
on thick paper, **$2** per hundred ; on card
board, **$5** per hundred.

Temperance Chimes. Price,
in paper, **30** cts. ; board covers,   **35**
Board covers, per doz., - - **4 00**
Paper covers, per doz., - - **3 40**

Temperance Hymn-Book.
Price, paper covers, **12** cts. each ; **$10**
per 100. Board covers, **15** cts. each ;
per 100, - - - - - **13 00**

TWENTY-FOUR PAGE PAMPHLETS.

Five Cents each ; Sixty Cents per Dozen.

Is Alcohol Food ? By Dr, F.
R. Lees.

Adulteration of Liquors.
By Rev. J. B. Dunn.

A High Fence of Fifteen
Bars. By the author of "Lunarius."

Bible Teetotalism. By Rev.
Peter Stryker.

Dramshops, Industry, and
Taxes. By A. Burwell.

Drinking Usages of Society.
By Bishop Alonzo Potter.

Duty of the Church toward
the Present Temperance Movement,
The. By Rev. Isaac J. Lansing.

Fruits of the Liquor Traf-
fic. By Sumner. Stebbins, M.D.

Gentle Woman Roused. By
Rev. E. P. Roe.

History and Mystery of a
Glass of Ale. By J. W. Kirton.

Is Alcohol a Necessary of
Life? By Prof. Henry Munroe.

Liquor Traffic, The—The
Fallacies of its Defenders. By Rev. E.
G. Read.

Medicinal   Drinking. By
Rev. John Kirk.

Physiological Action of
Alcohol. By Prof. Henry Munroe.

Son of My Friend, The. By
T. S. Arthur.

Stimulants for Women. By
Dr. James Edmunds, M.D.

Throne of Iniquity, The.
By Rev. A. Barnes.

Will the Coming Man Drink
Wine? By James Parton, Esq.

Woman's Crusade, The—A
Novel Temperance Movement. By Dr.
D. H. Mann.

TEMPERANCE SERMONS.
Fifteen Cents Each.

The National Temperance Society have published a series of Sermons in pamphlet form upon various phases of the temperance question, by some of the leading clergymen in America. Bound in one volume in cloth, $1 50.

1. **Common Sense for Young Men.** By Rev. Henry Ward Beecher.

2. **Moral Duty of Total Abstinence.** By Rev. T. L. Cuyler.

3. **The Evil Beast.** By Rev. T. De Witt Talmage.

4. **The Good Samaritan.** By Rev. J. B. Dunn.

5. **Self-Denial: a Duty and a Pleasure.** By Rev. J. P. Newman, D.D.

6. **The Church and Temperance.** By John W. Mears, D D, Professor of Hamilton College, New York.

7. **Active Pity of a Queen.** By Rev. John Hall, D.D.

8. **Temperance and the Pulpit.** By Rev. C. D. Foss, D.D.

9. **The Evil of Intemperance.** By Rev. J. Romeyn Berry.

10. **Liberty and Love.** By Rev. Henry Ward Beecher.

11. **The Wine and the Word.** By Rev. Herrick Johnson.

12. **Strange Children.** By Rev. Peter Stryker.

13. **The Impeachment and Punishment of Alcohol.** By Rev. C. H. Fowler.

14. **Drinking for Health.** By Rev. H. C. Fish.

15. **Scientific Certainties** (not Opinions) about Alcohol. By Rev. H. W. Warren.

16. **My Name is Legion.** By Rev. Stephen H. Tyng, D.D.

17. **The Christian Serving** his Generation. By Rev. Wm. M. Taylor, A.M.

TEMPERANCE TRACTS.

The National Temperance Society publish a series of tracts, among which are 190 12mo tracts, from one to twelve pages each, 72 18mo Illustrated Children's Tracts, all of which are put up in neat packets. Price 25 cents each.

Sixteen Temperance Leaflets, envelope size, in packets, 10 cents each.

LITHOGRAPHS AND POSTERS.

The Second Declaration of Independence. Size 12 x 19 inches. Per 100, - - - - - **3 00**

Five Steps in Drinking, 15

An Honest Rumseller's Advertisement. Per 100, - **1 00**

The Total Abstainer's Daily Witness and Bible Verdict, **75**

BAND OF HOPE SUPPLIES.

Band of Hope Manual. Per dozen, - - - **$0 60**

Temperance Catechism. Per dozen, - - - **60**

Band of Hope Melodies. Paper, - - - - **10**

Band of Hope Badge, Enamelled, $1 25 per dozen; 12 cts. singly. Plain, $1 per dozen; 10 cts. singly. Silver and Enamelled, each, - - **50**

National Temperance Orator, - - - - **1 00**

Ripples of Song. Paper covers, 15 cts.; per 100, $12. Board covers, 20 cts.; per 100, - - **18 00**

Juvenile Temperance Speaker. - - - **25**

Illuminated Pledge Card. Per hundred, - - - **2 00**

Temperance Medal. 10 cts. each; per dozen. - - - **$1 00**

Temperance Exercise. **10**

Illuminated Temperance Cards. Set of ten, - - **35**

Juvenile Temperance Pledges. Per hundred, - - **3 00**

Certificates of Membership. Per hundred, - - - **3 00**

Band of Hope Certificate and Pledge Combined (in colors). Per hundred, - - - - **4 00**

Temperance Lesson Leaves. Nos. 1, 2, 3, each 8 pp. Per 100, - - - - **1 00**

The Temperance Speaker. **75**

Catechism on Alcohol. By Miss Julia Colman. Per dez., **60**

TEMPERANCE PLEDGES.

1. Sunday - school Pledge, 20x28 inches, in colors, each, **$0 25**

2. National Pledge, 20 x 28 inches, in colors, each, - **25**

3. Family Pledge, 20 x 14 inches, each, - - - **30**

4. Family Pledge, 13½ x 10½ inches, per 100, - - **2 00**

5. National Pledges, for circulation at public meetings, per 100, **50**

6. Children's Illustrated Pledge, 9½ x 6 inches, per 100, **3 00**

7. Children's Illustrated Pledge, not including tobacco, and Certificate combined, 12 x 9½ inches, in colors, per 100, **4 00**

8. Children's Illustrated Certificate of Membership, 9½ x 6 inches, per 100, - - **3 00**

9. Children's Band of Hope Pledge, which includes tobacco and profanity, and Certificate combined, 12 x 9½ inches, in colors, per 100, - - - - **$4 00**

10. Pocket Pledge-Book, with space for 80 names, - **10**

11. Sunday-school Pledge-Book, space for 1,000 names, **1 50**

12. National Temperance Pledge-Book, space for 1,000 names, - - - **1 50**

13. Temperance Pledge-Card, 3½ x 5 inches, in colors, per 100, - - - - **1 00**

14. Illuminated Pledge-Card, per 100, - - **2 00**

Druggists', Property-Holders', Grocers', Dealers', Physicians', and Citizens' Pledges, per 100, - - **75**

TEMPERANCE DIALOGUES.

Trial and Condemnation of Judas Woemaker. 15 cents. Per dozen, - - - - **$1 50**

The First Glass; or, The Power of Woman's Influence; and

The Young Teetotaler; or, Saved at Last. 15 cents for both. Per dozen, - - - - **1 50**

Reclaimed; or, The Danger of Moderate Drinking. 10 cents. Per dozen, - - - - **1 00**

Marry No Man if He Drinks. 10 cents. Per dozen, **1 00**

Which Will You Choose? 36 pages. By Miss M. D. Chellis. 15 cents. Per dozen, - - - **$1 50**

Wine as a Medicine. 10 cents. Per dozen, - - - **1 00**

The Stumbling Block. 10 cents. Per dozen, - - - **1 00**

Aunt Dinah's Pledge. Dramatized from the Book, - - **15**

The Temperance Doctor. Dramatized from the Book, - **15**

Shall I Marry a Moderate Drinker? 10 cents. Per dozen, **1 00**

THE YOUTH'S TEMPERANCE BANNER.

The National Temperance Society and Publication House publish a beautifully-illustrated four-page monthly paper for children and youths, Sabbath-schools, and juvenile temperance organizations. Each number contains several choice engravings, a piece of music, and a great variety of articles from the pens of the best writers for children in America.

Its object is to make the temperance work and education a part of the religious culture and training of the Sabbath-school and family circle, that the children may be early taught to shun the intoxicating cup, and walk in the path of truth, soberness, and righteousness.

The following are some of the writers for THE BANNER: Mrs. J. P. Ballard (Kruna), Miss M. D Chellis, Mrs. Nellie H. Bradley, Rev. Wm. M. Thayer, Edward Carswell, Geo W. Bungay, J H. Kellogg, Mrs. J. E. McConaughy, Mrs. M. A. Dennison, Mrs. E. J. Richmond, Rev. S. B. S. Bissell, Rev. Alfred Taylor, Mrs. M. A. Kidder, etc., etc.

THE BANNER has already been welcomed into thousands of Sabbath-schools of all denominations as the only youth's temperance paper published for Sabbath-schools.

Terms, cash in advance, including postage:

Single copy, one year, - -	**$0 35**	Thirty copies, to one address,	**$4 05**
Eight copies, to one address, -	**1 08**	Forty " " " -	**5 40**
Ten " " " -	**1 35**	Fifty " " " -	**6 75**
Fifteen " " " -	**2 03**	One hundred copies, to one	
Twenty " " " -	**2 70**	address, - - - -	**13 00**

We trust the friends of temperance and Sunday-schools will make the effort to introduce THE BANNER into every Sunday-school in their midst, as the price at which it is published—which does not cover the cost of paper and printing—prevents the sending of agents to introduce it.